A NEW WAYS to HEALTH Book

Harriet Harvey, Editor

BOOKS BY SHERRY SUIB COHEN

No Naughty Cats
(with Dr. Debra Pirotin)

Diabetes
(with Lee Ducat)

Cristina Ferrare's Style
(with Cristina Ferrare)

Quizzical Pursuits

Making It Big
(with Jean DuCoffe)

The Looks Men Love
(with Vincent Roppatte)

About Face
(with Jeffrey Bruce)

Southern Beauty
(with Kylene Barker Brandon, Miss America 1979)

Tough Gazoobies on That!

THE MAGIC OF TOUCH

SHERRY SUIB COHEN

PERENNIAL LIBRARY

HARPER & ROW, PUBLISHERS, New York
Cambridge, Philadelphia, San Francisco, Washington
London, Mexico City, São Paulo
Singapore, Sydney

Grateful acknowledgment is made for permission to reprint:

Excerpts from *Touching for Pleasure* by Adele P. Kennedy and Susan Dean. Copyright © 1986 by Chatsworth Press. Reprinted by permission. All rights reserved.

Excerpts from *The Naked Ape* by Desmond Morris. Copyright © 1967 by Desmond Morris. Reprinted by permission of McGraw-Hill Book Company.

Excerpt from *Body Language* by Allan Pease. Copyright © 1981 by Allan Pease. Reprinted by permission of Sheldon Press.

Excerpts from *Encounters with Qi, Exploring Chinese Medicine* by David Eisenberg, M.D., with Thomas Lee Wright. Copyright © 1985 by David Eisenberg with Thomas Lee Wright. Reprinted by permission of W. W. Norton & Co., Inc.

Excerpts from *Love, Medicine and Miracles* by Bernie S. Siegel, M.D. Copyright © 1986 by B. H. Siegel, S. Korman, and A. Schiff, Trustees of the Bernard S. Siegel, M.D., Children's Trust. Reprinted by permission of Harper & Row, Publishers, Inc.

Special thanks to Ashley Montagu for his classic book on the subject: *Touching: The Human Significance of the Skin* (Third Edition). Copyright © 1971, 1978, 1986 by Ashley Montagu. Published by Harper & Row, Publishers, Inc.

A hardcover edition of *The Magic of Touch* is published by Harper & Row, Publishers, Inc.

FIRST PERENNIAL LIBRARY EDITION published 1988.

Copyeditor: Margaret Cheney
Designer: C. Linda Dingler
Indexer: Maro Riofrancos

Library of Congress Cataloging-in-Publication Data

Cohen, Sherry Suib.
  The magic of touch.

  "Perennial Library."
  (New ways to health)
  Bibliography: p.
  Includes index.
  1. Touch—Therapeutic use. 2. Touch—Psychological aspects. I. Title. II. Series.
RZ999. C57   1988   615.8′22   86-45649
ISBN 0-06-091457-2 (pbk.)

---

88 89 90 91 92 MPC 10 9 8 7 6 5 4 3 2 1

DEDICATED TO: THE REAL LARRYS

Larry Ashmead   "An editor cannot always act as he would prefer." HENRIK IBSEN

WRONG. The real Larry always acts as he would prefer and that is with wit and originality, gentleness and generosity, and love. Also a definite speck of lunacy.

Larry Cohen   whose touch I cherish.

CONTENTS

Acknowledgments   ix
New Ways to Health   xi
Foreword   xv

INTRODUCTION: What's Touch All About?   1

CHAPTER ONE:
THE ANTHROPOLOGICAL TOUCH: Touch Around the World and Through
the Ages   8

CHAPTER TWO:
LOVING TOUCHES   20

   Birthtouch   20
   Touch and Babies   23
   And Then They Grow (Touch and Young People)   34
   Do You Love Me for My Body or My Mind? (Touch and Sex)   39
   Don't Stop Now (Touch and Growing Older)   49
   The Last Touch (Touch and Dying)   53

CHAPTER THREE:
HEALING: The Historical Healer   57

CHAPTER FOUR:
HEALING: The Touch of Energy   63

   Therapeutic Touch   66
   Acupuncture and Acupressure   75
   Chinese Massage   83
   Shiatsu   84
   Reiki   87
   Polarity Therapy   91
   Reichian Massage   93
   Bioenergetics   96

CHAPTER FIVE:
HEALING: The Touch of Manipulation   98

   The Rub and the Crack: Manipulative Medicine   98
   Rolfing (Structural Integration)   100

The Alexander Technique   102
The Feldenkrais Technique   104
Osteopathy   107
Chiropractic   110
Applied Kinesiology   113
Touch for Health   118
Swedish Massage   121
Reflexology (or Zone Therapy)   123
Medical Massage and Hospital Touch   126
The Self-Massage   132

CHAPTER SIX:
FAITH HEALING   **135**

CHAPTER SEVEN:
TOUCH TACTICS   **143**

The Power Play (Touch as Silent Force)   143
The Convincer (Touch Talks You Into It)   152
The Communicator (Touch Gives Your Message)   156
The Word (Touch *Becomes* Language)   163

CHAPTER EIGHT:
FINISHING TOUCHES   **166**

A Touch of Spot   166
Touch Taboos   172
Turn-offs and Turn-ons   180

Resources   **187**
Bibliography   **195**
Index   **199**

ACKNOWLEDGMENTS

To Harriet Harvey, whose vision gave life to this book and whose intelligence inspired clarity.

To my agent,
   Connie Clausen   Knowing I love'd my books, he furnished me
                                From mine own library with volumes that
                                I prize above my dukedom.
                                       William Shakespeare, *The Tempest*

To Adam, Jennifer, Steve, Jane, David, and Lee, who touch me with love and friendship every day of my life. Well, almost every day. To Sallie W. Coolidge, an editor who epitomizes skill, warmth and support, and to Ann Martin-Leff, her fine assistant.

Many others touched this book with a helping hand and I'm especially grateful to Richard Grossman from the Residency Program in Social Medicine and the Department of Family Medicine at New York's Montefiore Hospital; Joan Mowat Erikson, Mamaroneck Public Library; Cathleen A. Fanslow, R.N., M.A.; Dr. Richard S. Heslin; Evelyn Lauder; Dolores Krieger, R.N., Ph.D.; Allen Forbes, Brown University anthropologist; Dr. Rufus Peebles, Cambridge psychoanalyst; Trish Turk, typist extraordinaire.

NEW WAYS
TO HEALTH

This book is the first of a new Harper & Row series, *New Ways to Health*, which proposes to build a bridge between the theories and therapies of traditional Western medicine and what is known popularly as "alternative medicine." Along with many physicians and health practitioners, we no longer consider much of "alternative medicine" as alternative, but rather complementary or adjunctive to our Western scientific methods. Many of these new or age-old alternative therapies are extraordinarily valuable and are badly needed to serve our day-to-day health needs as fully as possible. The two approaches should be combined in what Richard Grossman of Montefiore Hospital, New York, calls a new "ecumenical medicine."

Western medical practices are now undergoing a major revolution on many fronts. Scientific medicine—especially in the area of genetics and molecular biology—is making quantum leaps. At the same time, a new field, with the tongue-twisting name of psychoneuroimmunology, is proving that the mind and body are so intimately connected that many diseases that we thought were "purely physical" or "purely mental" are, in fact, an interwoven admixture of the two—and must be treated as such. This seems to be particularly true of our big unsolved chronic disorders: heart disease, gastrointestinal disorders, and cancer.

Simultaneously, preventive medicine is burgeoning, as we discover that a change in our lifestyle habits—nutrition, exercise, stress management—will not only ward off disease, but bring us more vital health.

The ancient and venerable arts of Chinese and Indian medicines —and those of many other cultures—have never made the error of separating the mind, body, and spirit in their therapies. Moreover, much of Eastern medicine is preventive and focuses on nutrition and exercise. For these reasons alone, it is invaluable for us to include many age-tested practices—such as yogic meditation, exercises from the mar-

tial arts, acupuncture, breathing exercises, and herbalism—within our
own pharmacoepia. At the same time, we must revive and update some
of our own ancient Western healing arts—such as the power of touch
—which were thrust aside by the scientific revolutions of the last two
centuries.

Books for the *New Ways to Health* series will approach common
illnesses and health concerns in terms of effective therapies from the
medicines of many cultures and offer new and valid health practices to
the public as soon as they become available.

Members of the medical advisory board for *New Ways to Health*
include: *Philip R. Lee, M.D.,* Professor of Social Medicine and Director
of the Institute for Health Policy Studies at the University of California,
San Francisco. Heavily involved with the nation's health care and
health delivery systems, Dr. Lee was Chancellor of UCSF from 1969 to
1972 and was the Assistant Secretary of Health and Scientific Affairs,
Department of Health, Education, and Welfare from 1965 to 1969. In
1985, Dr. Lee was appointed President of the Health Commission for
the City and County of San Francisco and, in 1986, was appointed
Chairman of the Physician Payment Review Commission of the Con-
gressional Office of Technology Assessment. He is the author of over
100 articles on health and medicine, the coauthor of three major books
and a frequent advisor to federal health policy makers.

Andrew Weil, M.D., Lecturer in the Division of Social Perspectives
in Medicine at the University of Arizona College of Medicine; author
of *Health and Healing* and other health-related books.

Richard Grossman, Director of the Project for Health in Medicine
at Montefiore Hospital, and Assistant Professor of Epidemiology and
Social Medicine at the Albert Einstein College of Medicine, New York,
and author of the recently published *The Other Medicines.*

Dolores Krieger, R.N., Ph.D., Professor of Nursing at New York
University and developer of the healing technique, Therapeutic Touch.

Lawrence Le Shan, Ph.D., an expert on psychological factors in
cancer, and current President of the Association for Humanistic Psychology; author of *You Can Fight for Your Life* and other books on
psychology and health.

Clarence E. Pearson, M.P.H., Director of the Metropolitan Life
Insurance Company's Health and Safety Education Division, and President of the National Center for Health Education.

M. Barry Flint, Executive Director, and *Harris Dienstfry,* Director of Publications, for the Institute for the Advancement of Health, New York, an organization which acts as a clearinghouse for new research in mind-body research and health care.

Harriet Harvey
Editor, *New Ways to Health*

FOREWORD

Touch is one of our most precious human resources—powerful as communication and powerful as therapy, and sometimes vital to the support of life itself. Yet, for the better part of two centuries, we have all but forgotten its many functions, sometimes with disastrous consequences.

In the 1920s, foundling hospitals in the United States were reporting a death rate of nearly 100 percent for babies under a year old. The physicians couldn't figure out what was wrong. The hospitals were antiseptic, the infant formulas nutritious and perfectly prepared. Then, through a chance observation on a trip to Germany, one of our pediatricians discovered that a children's ward in Berlin employed an elderly woman to come in and cuddle the ill babies—and most of them got well. When the visiting doctor brought the word back to the United States, his fellow pediatricians had a hard time believing that so simple an interaction could be so effective. But gradually, through experimenting, they came to realize that human touch was vital to infants and, when they allowed holding and cuddling in the hospital, the death rate actually reversed itself. Sherry Suib Cohen, the author of this book, tells the full story of this incident in Chapter Two.

Touch for health and for healing was not always so ignored. Until the scientific age, in the West it was one of the foremost therapies in a healer's repertoire and it has a very long history. Cave paintings in the Pyrenees attest that some fifteen thousand years ago people were practicing what later came to be known as the laying-on of hands. The Ebers Papyrus describes a similar Egyptian treatment with the use of touch that dated back to 1552 B.C. Throughout the Far East, ancient traditions of healing by hands have been handed down generation after generation from teacher to pupil and are still in use today in China, Tibet, and India. Healing by touch is described in both the Judeo-Christian Old and New Testaments, as well as in writings about the Roman emperors

Vespasian and Hadrian. It was as much a ritual in the early Christian period as prayer. After the Roman Catholic Church gave up the practice, the kings of France and England took over and administered "the Royal Touch," which was considered particularly effective in curing goiter and scrofula.

Church histories from the Middle Ages recount innumerable cases of healing by laying-on of hands. It is sad to note, however, that healing outside the Church was looked upon with suspicion and thought to be witchcraft, imprecations of the devil, or, at best, mere nonsense.

In the late eighteenth century, Franz Anton Mesmer was a highly fashionable medical doctor in Paris. He practiced—very successfully—a form of laying-on of hands and claimed that its healing properties resulted from a transfer of "magnetic fluid" from his body to that of his patient. This concept is somewhat akin to the Eastern idea that the human body has energies—called Prana or Qi (Ch'i)—which move through nonphysical pathways (meridians) and that this energy can be buttressed or modulated through healing techniques such as acupuncture or Chinese massage or through self-healing practices like Yoga or other forms of meditation. Mesmer, however, was discredited by a blue-ribbon commission appointed by the king of France and, after that, the idea that healing energy could be transferred from one person to another disappeared in the scientific circles of the West for over one hundred years. The laying-on of hands was dismissed as nothing more than religious superstition.

Touch was too simple a process to appeal to the scientific age with its adulation of things mechanical, complex, and synthetic. Only now, as we are discovering that we can't cure all our ills with machines and pills, is touch—and a unified approach to body, mind, and spirit—coming back into its own. We are also experiencing the influence of Eastern healing practices—which never discarded touch therapies nor segregated body, mind, and spirit in their search for health. Through ideas from other cultures coupled with new scientific methods, we are beginning to broaden our view and appreciation of the many powers of touch.

With touch, we have the extraordinary opportunity to recapture a simple but elegant mode of healing and mate it with the rigor and power of contemporary science. An understanding of the interface of touch and the human condition has only just begun.

In my own work with touch, particularly with therapeutic touch, it becomes clear that the therapeutic use of hands is a universal human potential. It is not a gift restricted to priests and kings or other specially chosen people. With little exception, anyone can learn. Undoubtedly, there are some people who are born with a finer ability to perceive subtle human energies. My teacher, Dora Kunz, was born with that capability, but she has trained herself so that her ability to discern has developed into a finely tuned, controlled instrument. However, through compassion, sensitivity, and practice, anyone motivated to heal or help others can do much to relieve pain and illness.

Some of the therapies in this book are complex and it often takes many years to become proficient in them. Others are comparatively simple. The basics of the major technique I use—Therapeutic Touch—can be learned in a few weekend workshops, although one becomes more skilled and understanding of the process as one continues to practice it. What one learns from all therapies is a greater sensitivity to the needs of fellow human beings, and to your own needs as well. You become more highly tuned to subtle energy differences within the body, both of the client and yourself, and they become meaningful to you.

A most distinctive feature of this book is that the author addresses the need for simple, everyday touch as sensitively as she describes the often highly refined touch therapies. All kinds of compassionate humane touch are needed to counter our rigid, machine-dominated society. Babies and small children particularly need to be hugged and held —and so do the elderly. In actuality, this may be true for all of us. We do know that every bit of practice of simple touch or of a healing technique will enrich one's life; that is its power, that is the magic.

Dolores Krieger, R.N., Ph.D.
New York University

WHAT'S TOUCH ALL ABOUT?

Everyone, I remember, was looking up, necks craned, stretching to see. The air was cool and still, the room vast, the ceiling towered above. The voices of the other tourists were hushed.

I'd waited most of my life to see the Sistine Chapel, the glory of the Vatican fathers, and Michelangelo's masterpiece of The Creation, when God, according to the book of Genesis, "formed man . . . and breathed into his nostrils the breath of life." Now in Rome with my husband, I looked up eagerly.

But wait. Something was odd. I'd seen dozens of pictures of the frescoes but I'd never before noticed something that struck me now. As I peered intently at the forms in the dim-lit chapel, it was apparent that God was not breathing life into Adam's nostrils at all. He was *touching* life into his creation. With one finger, reaching across an abyss marked with the cracks of time and Buonarroti's peeling paint, God passed sentience and spirit into the still form of Adam.

It made sense. Then, as now, the idea thrills me.

Touch is, quite simply, the most fundamental of all the senses. Asleep or awake, it gives us knowledge and forms our perception of love, hate, shape, softness, hardness, thickness, hot and cold. In evolution, the sense of touch developed first and it is also the first sense to develop in the human body. Consider this: touch is born before we are born; when the human embryo is only the length of a nice-sized string-bean and before it develops eyes or ears, it is already reacting to touch. Before we meet the cold light of day, before we are technically "begun," our sense of touch is educated in our mothers' wombs. In some collective memory, we've all known how it feels to be nudged by the pliant walls of the womb, to be bathed by the warm, buoyant amniotic waters.

The process of birth itself is a touch act. The voyage down the uterine canal is marked by pre-birth squeezes, which invigorate the untried respiratory and digestive systems of the emerging infant. The intense contact provided by massive uterine contractions, and the immediate touches of parents and medical attendants—it is in these earliest moments that the sense of touch is reinforced. The taste for touching and being touched is profound.

What's touch all about? The average human body is covered with about eighteen square feet of skin, pressed thin as baklava pastry and weighing in at about eight pounds. This covering is studded with about five million tiny nerve endings, which are touch receptors. On the tip of only one finger, there are seven hundred of these receptors for every two millimeters of skin, so when a mother touches her baby, or a lover his beloved, all kinds of extraordinary reactions are set in motion.

Choose any square inch of skin: in it can be found about fifteen feet of blood vessels and about seventy-two feet of nerves. Its thickness will change from about one-fiftieth of an inch on your eyelid to as much as one-third of an inch on the palms and soles of the feet. Three layers of skin fit snugly together. The top layer, the epidermis, is tough and horny and replaces itself every twenty-eight days, as dead skin cells flake off. The next two levels, the dermis, contain blood vessels, nerves, glands, and the fibroelastic connective tissue that gives the skin its ability to spring back and retain its shape after a touch.

Skin is a miraculous envelope. It covers us and protects us; it is waterproof and repairs itself. Specific nerve endings are activated by touch and send their particular sensory messages along the spinal cord to the brain.

"The skin is the largest of all our organ systems and, perhaps next to the brain, the most important of all our organs," says anthropologist Ashley Montagu, who wrote the seminal work *Touching, the Human Significance of the Skin*. "It should be thought of," he says, as our "external nervous system, an organ system which from its earliest differentiation remains in intimate association with the central nervous system."

For some unique people, like the child Benjamin, the skin is even more important than the brain. Benjamin is a very real little boy. Incredibly, he was born without a brain—a feat almost unheard of in medical literature. With only a brain stem, he lives and breathes. Benjamin has nothing, nothing at all, to take from the world or to give to it. He exists in a colorless, mindless vacuum all his own.

And then the magic is applied. When Benjamin is picked up and touched, he smiles. He laughs. He still cannot hear, taste, see, or think, but touch makes him smile and touch pulls him into the world.

They're all grand—hearing and smell and sight and taste, but touch —touch is elemental. There's no kidding around about touch; touch is a biological necessity without which the body as well as the soul falters. Science has shown that babies can live, as Benjamin does, without seeing, hearing, tasting, or smelling, but, if they are not touched, they wither away and die.

Thus, from those earliest moments when we begin to feel inside the womb, touch becomes part of all our days and nights, our education, our ritual. It nourishes, entices, influences, empowers.

Most important, it heals. After years of neglect from the medical profession, many health practitioners are once again recognizing touch as healer. Throughout history, physicians and nurses have measured extraordinary therapeutic benefits from touch. But, every now and then, a time period occurs when touch falls out of favor, as it has in recent Western medicine. Inevitably, the pendulum swings back and we come back to touch. Today, therapies as old as the hills are once again being rediscovered by traditionalists. The touch of well-being, knocking around throughout recorded history, despite orthodox medicine—and sometimes because of orthodox medicine—continues to be an irresistible healing alternative to whatever is currently de rigueur.

It's interesting: even when touch has not been an accepted means of treatment, there have always been mothers who cured a hurting place with a kiss or cooled a raging fever with a loving hand. If physicians have waffled from era to era about touch's efficacy, it was never an idea that went out of favor in the secret, dark places of the heart.

The eighties are witnessing the mind-body revolution as biological scientists are beginning to demonstrate that there are very intimate connections between the brain, the nervous system, and the immune system, and that all the functions of the mind—thinking, believing, feeling, imagining—have a profound effect on our physiological health and well-being.

The laying-on of hands is taking on renewed importance for healing. Not only is it being revived in many religions—particularly among Catholics and Episcopalians—but well-informed doctors and nurses are learning its techniques. The rash of new massage parlors are not, by a long shot, all fronts for sexual play; many practice highly developed tactile healing arts, from ancient Japanese Shiatsu to newly invented therapies such as Rolfing.

Touch is the most tangible of senses, and still varying attitudes make it the most mystical. Perhaps nowhere has touch been so tangible *and* mystical as in the way the world's religions see it. From the taking of the sacrament to baptism, from the kissing of the Pope's ring to the fourteen of Jesus's miracles that relied on touch, Christianity is literally infused with touch. Go to a synagogue in any ancient or modern section of any town and see the way the Jews reach out to touch the Torah as it's carried triumphantly down the aisle. Go to the United Methodist Church in suburban Kansas City and check out the sign at the main entrance which reads, "For Healing Use the Back Door of the Church." Perhaps it isn't yet accepted in mainstream Kansas City, but at least healing is available with a touch from the Methodist minister at the back door.

Go to a Buddhist shrine, an Islamic mosque, a Taoist temple and see the way religious people touch their consecrated garments, their icons, their religious tokens, each other in both real and symbolic ways.

Listen to the way D. H. Lawrence remembered the way one Italian church awakened his touch sensibility:

> I went into the Church. It was very dark and impregnated with centuries of incense. It affected me like the lair of some enormous creature. My senses were roused, they sprang awake in the hot spiced darkness. My skin was expectant, as if it expected some contact, some embrace, as if it were aware of the contiguity of the physical world, the physical contact with the darkness and the heavy suggestive substance of the enclosure. It was a thick, fierce darkness of the senses.

"Expectant" skin, writes Lawrence. It is a telling concept. We all, piglets, puppies, men, and monkeys have expectant skin.

Touch is communication. Touch talks. It speaks in voices that are louder and clearer than any sound that ever sprang from a diaphragm. Touch tells people what to do, tells them what *you* think, tells them all about you. Touch changes languages and psyches. Touch is a politician. It is also a love-maker with a vibrant libido.

Throughout history cultures have differed in their predilections to touch, but there has never been a people who has not touched at all— even though, as we shall see, there have been some that have touched very little and have turned sour for the lack of it. As long as record keeping has existed, there have been people who etched on stone walls, quilled on delicate parchment, or rolled out with their ballpoint pens the message that their tribe was one of warm-blooded creatures who, with a simple hug or handclasp, said to each other, "I'm here, you're not alone."

Touch is also a political animal, and the politics of tactual assertion are well known in boardrooms and bedrooms or wherever people claim power.

Touch is sensual, erotic, joyful in its sexuality. An enlightened view of touch teaches that the entire body is an erogenous zone and that sexual pleasure should not be limited to a race for the elusive orgasm. Cuddling, massaging, stroking, tickling when developed to their fullest art forms can make the sexual experience exquisitely satisfying.

In a world where we rely on Hallmark cards and machine-answered telephones to give our messages, often one receives faulty signals because words can mask true feelings, and messages that you can buy for a dollar are not terribly dependable. But just try to fool touch! It makes our skin warm, glow, glisten, blush, hurt, tense, or tremble with the clarity of its messages. When optical illusions fool our eye and brain, and we finally learn that spoken and written words can't be totally trusted, it's touch that comes poking through to tell the truth.

Touch is a telephone to the world.

· Your business associate says he's not stealing your money. His clammy, cold handshake tells you the opposite.
· The patient is comatose, feels nothing and is as good as dead, say the doctors. You stroke and hold her, and when the nurse measures her hemoglobin level, it's grown higher! After the patient recovers, she tells how she clearly remembers loving touches pulling her back.
· The boss says he's all for women's rights. Then he pats his secretary on her hip and you know he's pulling rank in an anti-woman manner.
· A famous television star is suffering dreadfully from a kidney stone. Nothing the doctors do helps. He's scheduled for surgery. Then he gets a foot massage. In six hours, he passes the stone.
· Secret handshakes provide entry into old-boy and little-boy clubs all over the world.
· Your lover lightly runs a hand along your arm—and your body quivers with expectation and excitement.

In short, the magical power of touch is a tool that lets us change the world. It has been referred to as the common touch but it is most uncommon. Down through the centuries, down through the psyches, down through the sensibilities of every Adam and Eve, every Napoleon, Plato, Shakespeare, Queen Elizabeth, Yogi Berra, Ronald Reagan, Gloria Steinem, every Joe Blow who has ever walked this earth a mighty hunger for the uncommon touch has existed. It is ancient, eternal, omnipresent. Everyone has it. I have it, too.

As a wife, mother, teacher and writer, I am fascinated by touch.

Standing before the Michelangelo frescoes in Rome, I felt a kinship with the artist who painted creation as a touch. One morning, in New York City, I found a painting hanging in the Whitney Museum of Art. It is by the contemporary artist Alex Katz. He has rendered a group of couples, very chic, very *now,* and each is touching the other and passing messages through those touches that are as disparate as day and night. One person touches his partner with love, another with comfort, another seems to be nurturing, yet another is hateful and clawlike in her touch. Amazing. Two artists, centuries and styles divided but unified in their respect for the magical power of touch. Both have pulled me into their visions.

When I was approached to write a book about touch, I was delighted. I would have a chance to learn more about the specific effects of touch. In addition to conducting the usual research through books and interviews with experts, I decided to experience personally many of the touch therapies so that I could describe them firsthand.

I started out a skeptic, for, in spite of my love of touch, I wasn't sure that its touted ability to heal didn't emanate from old wives' tales or myths from the California touchy-feely craze. I was a believer in the "regular docs"—the medical establishment as we know it. I still am. But since I've experienced some healing magic, I've simply added some of these new/ancient touch therapies to my pharmacopoeia. Even within the medical establishment, many of the touch therapies—new or old— are gaining recognition—particularly as mind-body interactions are now being proved increasingly potent in promoting or diminishing our health.

Dr. Bernie S. Siegel, a cancer surgeon at the Yale Medical School, writes in his new book, *Love, Medicine and Miracles,* "I personally feel that we do have 'live' and 'die' mechanisms within us. Other doctors' scientific research and my own day-to-day clinical experience have convinced me that the state of the mind changes the state of the body by working through the central nervous system, the endocrine system and the immune system. Peace of mind sends the body a 'live' message while depression, fear and unresolved conflict give it a 'die' message."

After my experience with many of the touch therapies, I have come to believe that touch gives a 'live' message—sometimes when everyone is saying there's little hope for life. More often it is not as dramatic as live or die; sometimes it's just a matter of getting the pain from the tennis elbow to cease and desist when the cortisone shot doesn't work and the doctor has no other solutions in his little black bag.

In the little black bag of touch therapies, there are many possible approaches and they're well worth trying. My personal reaction to each therapy is not meant as a recommendation; it is merely a description of my own perceptions. As you read, trust your own perceptions and reactions. More often than not, they will be a sound clue to what will work best for you.

As you will see, at the end of each section describing a specific therapy, I have made a list of what that particular treatment is best used for. Under "Resources" (page 187) you will find organizations through which you can locate therapists in your locale.

But, before we move on to the specific touch therapies, let's look at the magic of touch in different cultures around the world.

CHAPTER ONE

THE ANTHROPOLOGICAL TOUCH:
Touch Around the World and Through the Ages

Even a casual look at the anthropology of touch has to start with the primate culture that preceded man, and that means Harry's monkeys. Anyone who knows anything at all about touch knows about those monkeys. Harry himself had no idea how famous they'd be.

Professor Harry Harlow was a scientist at the Wisconsin Regional Primate Center in the late fifties and his work involved study with monkey families. One day, he noticed that the baby monkeys in the laboratory were spending an inordinate amount of time with the folded gauzy diapers that their keepers used to cover the cage floors because it had been shown that baby monkeys rarely survived on the otherwise bare, wire-mesh floors of the cages. Whenever the keepers attempted to remove the diapers to wash them, they had an onslaught of angry, screaming, clawing baby monkeys on their hands. This would come as no surprise to any mother who attempts to launder her toddler's security blanket. At any rate, Dr. Harlow decided to build two "surrogate" mothers in the monkey cage. Both would be built of wire mesh but one "mother" would be covered with touchable terry cloth and she would be lit with a bulb that would give off delicious heat, similar to a real mother's body warmth. This surrogate would be, Harry said, "soft, warm and tender, a mother with infinite patience, a mother available twenty-four hours a day, a mother that never scolded her infant and never struck or bit her baby in anger." Pretty appealing.

The other mother would have its appealing virtues, as well. It would be hard to beat food as a virtue and, while the terry-cloth mother gave comfort and no sustenance, the plain wire mother would have a nipple to dispense milk.

No contest. Hour after hour, the babies clung to the mother who gave the tactile comfort and only visited the "milk" mother to avoid starvation. To Harry's great amazement, the touch of security took precedence over food. It was amazing. Touch over sustenance? Yes. Monkeys do not live by bread alone. Neither do people. It's important to note, however, that *all* Harry's monkeys eventually had difficulty with social interaction and all of them proved sexually retarded. Mating, which ought to have come naturally, simply didn't. Surrogate fake wire mothers, comforting terry cloth or not, still did not provide the personal touch interaction the monkeys needed; the inadequate social experience made them dreadfully neurotic.

You'd think that Harry's lesson would have been carried over a few years, but no. Chimpanzees, raised for research, have traditionally been separated from their offspring, the infants being carted off to sterile and controlled laboratories at birth. What's been happening? Chimp social flops have been flooding the laboratories. Socially deprived of parent-child interaction and touch, the animals never learn how to copulate or care for infants. Females scream at the approach of males and, when screaming fails to take the male's mind off cuddling, they groom him furiously until he forgets what he came for.

At the University of Colorado Medical Center, Professor Martin Reite studied the intricate and massive effects that the deprivation of a mother's touch means to baby primates. Babies, separated from touch, show profound grief, huddle in corners, and act in classically depressed manners. If the mother is returned, after a few days, the obvious symptoms disappear—but physical injury remains. The babies are more susceptible to disease and a broad range of bodily weaknesses. Body temperature, heart rate, brain-wave and sleep patterns are disturbed. Scientists see disturbances in immune function, and psychological problems range from difficulty in cooperation to out-and-out violence and social isolation. If monkeys are deprived of touch for just a short time, then returned to their mothers, the problems are ameliorated but never completely disappear. If they're touch-isolated for long periods of time, then returned, very serious problems are less likely to "get better." The parallels between monkey and human behavior are immense, think scientists.

So what to do? Researchers rely on chimpanzees for study into human problems, which include finding cures for hepatitis, AIDS, and other diseases that involve blood sampling and liver biopsies (which, incidentally, do not harm the chimps).

The message of touch is not lost on the most forward-thinking of today's scientists. Taking a page from Professor Harlow's book, Dr. Jan Moor-Jankowski of New York University's Laboratory for Experimental Medicine is currently directing a program where newborns are kept with their mothers and other family members for a year and a half after birth and, shades of Harry's experiments, the studies so far are proving that the early experiences of mother bonding and family play involving tactile contact are making happier, healthier, sexually interested chimps. As they learn to touch, they learn to live.

Humans touch as instinctively as they breathe. The first formal touch was undoubtedly a handshake between two cavemen, say anthropologists. As the earliest men met, they'd first hold their hands high in the air, palms exposed to show each other that no clubs were on the ready. Drawings found in neolithic caves portray cavemen with palms coming closer and closer through the years, as more and more good will was expended; eventually, palms in the air evolved to clasped palms. It figures. Even cavemen had to cooperate and touching is the only sense that relies on cooperation. You can't give a touch without getting one right back. You can talk, listen, smell, see, and taste alone, but touch is a reciprocal act.

Pictorial evidence in cave paintings in the Pyrenees, showing people using touch as a restorative and influential power, goes back an estimated fifteen thousand years. In China, Egypt, and Thailand, the ancient traditions of touch healing are illustrated in early rock carvings, in painted papyrus, in the lessons still being handed down from teacher to student. Touch for bonding is as varied as an Arab feeling and smelling his guest to an Eskimo rubbing noses with his guest—each in a burst of loving friendship. For some Middle Easterners, denying a friend the touch of your breath (garlicky or hot as it may be) is to be ashamed of that friend. For Far Easterners, touch is seldom a public affair. Although Japanese literature is rich with the most erotic touchings imaginable, Japanese friends rarely touch outside the home but, on meeting, bow lower and lower and lower in a symbolic touch of respect. Greeks and Italians don't do much bowing, but subscribe to gusty embraces through every joy and crisis. American Indians and Russians who swaddle their babies do so from some atavistic memory that reminds them how comforting a tight, snug cloth embrace can be.

In the writings of both the old and the new Judeo-Christian scholars, touch was pictured as a healer and a bonder. The Pharaohs of ancient Egypt started and ended each day with a massage. In the first half of the thirteenth century, Holy Roman Emperor Frederick II conducted an experiment that had infants being touched only when they were bathed or suckled. The historian Salimbene recorded that all the babies died because they could not "live without the petting and joyful faces and loving words."

No one should knock joyful faces or loving words, but they don't hold a candle to touch.

Ashley Montagu tells us about the simple, uneducated Netsilik Eskimos who live on the Boothia Peninsula in the Canadian Arctic; a baby Netsilik is a real doll and Ashley Montagu says that touch is the reason. Netsilik mothers communicate with their babies almost entirely through their skins, since their infants spend most of their time naked, supported by slings, on the mothers' naked backs. When it's hungry, a Netsilik baby lets its mother know—not by crying but by sucking on her back. When it's tired, it sleeps on her back, skin to skin. When it has to urinate or defecate, it does so—right there on her back, which, incidentally, causes no great horror to either baby or mother, who cleans them both up in a totally accepting manner. Indeed, almost every single Netsilik Indian baby's needs are anticipated and handled through touch bonding between mother and child. Netsilik babies rarely cry. They grow to be warm, altruistic, and calm adults, even in the face of enormous stress (say, an angry polar bear).

The Mbuti hunter-gatherers, who live, as they have for centuries, in the Ituri forest of northeastern Zaïre, also don't have too many graduate-school degrees hanging in their mud huts. The Mbutis make a point of passing their newborn infants around to assorted family and friends, and the infants learn, early on, that multiple warm touches are even better than just one or two of the same. Mbutis are the happiest of hunter-gatherers. Mbuti males, by the way, delight in bodily contact throughout their adolescence, and this has nothing to do with homosexuality. They sleep together, as anthropologist Colin Turnbull puts it, "in a glorious bundle of young life, full of warmth and full of love."

The Kaingang tribe of Brazil's highlands do the same and they, says Jules Henry in his book *Jungle People,* "lie like cats absorbing the delicious stroking of adults. . . . Married and unmarried young men lie cheek by jowl, arms around one another, legs slung across bodies, for all the world like lovers in our own society."

In our American society, young men who secretly yearn for such touching tend to find it only in the contact sports that bring them black and blue marks as well as touching. Sad, actually, to consider this: wars are rarely fought between men who have shared a loving touch.

The Tasaday, a honey of a Philippine Stone Age tribe, which was discovered by modern-day anthropologists in the Mindanao jungle in 1966, use touch to express their unity in relationships among the aged, adult, and youthful members of the tribe. During stress and during pleasure, the entire tribe huddle together, touching, stroking, sharing body warmth and psyche strength.

By contrast, the Ugandan Ik is not such a honey of a tribe. Iks are characterized by their not very lovely habit of letting their infirm old or ill young fall by the wayside with a kick or laugh from a tribesperson if the member cannot keep up with the rest, for one reason or another. Iks hardly ever touch.

The Balinese, studied in depth by Margaret Mead, receive touch conditioning from earliest youth. A Bali couple would never, ever, opt for twin beds like many modern Western couples, who think of bed as a place to be alone (except for the purposes of making love). Middle class Americans, for the most part, spend their whole sleep life before adulthood alone in bed. A Bali baby, on the other hand, is constantly carried on the hip of a parent or older child until she's at least six months old; she's always in proximity with others, sleeping, dozing, playing, listening. The very purpose of clothing is to bind the baby to her mother. Balinese adolescents often fall asleep just leaning on each other. Bodily touch to a Balinese baby is as comfortable as Doctor Denton feet pajamas are to an American baby. And the Balinese are not a warlike nation, it might be observed.

Samoans express love by pressing one cheek to the cheek of a loved one, all the while taking staccato breaths that create "airjet" sounds from their noses. In fact, the Samoan verb "to doctor" is synonymous with "to rub"—and rub, and knead, and stroke the Samoans do. American Indians have always used a technique of light stroking to heal, sometimes even with no apparent physical contact as they only touch the air above the skin.

In Central Africa, the Azande tribe has individuals who are admired for their skill in reducing fractures by touch; the Azandes, like the Samoans, are also known for their skillful, light massages.

The natives of Tonga, reported eighteenth-century traveler William Mariner, used rubbing along with a doubtful technique that was unique; they'd employ little children to trample upon patients. It is easy to see the early origins of chiropractic and other manipulative therapies in many breeds of touches the world employs.

While touch and massage were often used in much the same ways we use them today, sometimes magic and fantasy were at the core of techniques. For instance, in the Solomon Islands in the early 1900s, the English anthropologist William Halse Rivers watched a massagelike activity taking place, and only upon asking did he find out that the massagers were getting rid of an imaginary octopus that had taken to napping in the body of a patient—or so the islanders believed.

Many cultures use touch as a way to prepare their littlest members for the world. They seem invariably to be the most peaceful cultures. Take swaddling: it is an ancient practice but often misunderstood. How can a baby express herself if she's all tied up? ask detractors of the practice. How can she kick and scream? Now, it appears, babies who are properly swaddled don't *want* to kick and scream, so comforting is their tactile bundling. The trick is to swaddle completely, which means a swaddled baby should be immobilized by his blankets, thereby getting the maximum of touch instead of being just restricted in movement, which makes him good and mad.

In !Kung territory, all the babies are plump, nurse at will, wear no clothes, and are in skin-to-skin contact with their mothers as they ride in soft leather slings at their mothers' sides. "The !Kung never seem to tire of their babies. They handle them, kiss them, dance with them and sing to them. The older children make playthings of the babies," writes Lorna Marshall, who lived among the Bushmen for over ten years. !Kung grow up to be awfully nice guys.

Montagu writes of the Mundugumor in the early thirties, a tribe that contrasts greatly to the !Kung. Mundugumor live by a river in New Guinea, and until recently had a hard time deciding on whether or not they wanted to keep their children. If those children could have talked, they would undoubtedly have reported they weren't sure their parents did them a favor by keeping them. Mundugumor infants were destined to spend their young lives in unpleasant touch, suspended high above their mothers' heads in scratchy, closed-in baskets. Body warmth and light of day were not available to be enjoyed. If the baby cried, no one touched it or rocked it—just scratched the outside of the straw basket, which made a grating sound. If a Mundugumor baby was really hungry and insistent upon food, its mother grudgingly suckled it—but standing up, no fooling around here, no enjoyment of touch! The moment dinner was over, it was back to the basket. Weaning was accomplished with shoves away from the breast and a few hearty slaps. It should surprise no one that the adult Mundugumor was a nasty, peevish, cannibalistic creature. Montagu notes that the Mundugumor of today, influenced by some of the kinder ways of other tribes, are very different from their parents and one can only feel pleased for the newest Mundugumor babes.

My husband has a sweet touch memory that involves death, oddly enough. When his grandmother was breathing her last few hours of oxygen, the doctor in attendance told my then sixteen-year-old husband-to-be to hold her in a sitting position in order to make her breathing easier. He did so, at first in trepidation, and then in a calm joy. Both of them felt gentled and succored at the mutual touch, and when the eighty-year-old died in the sixteen-year-old's arms, it was not frightening but inspiring.

It is interesting to note Montagu's description of the Atimelang tribe of the Netherlands East Indies island of Alor, who do the same thing as a matter of tribal duty. When someone is dying, it is the custom for one of the grown children or some other kinsman to hold the dying person in his lap. In recent years, the expert on death and dying Dr. Elisabeth Kübler-Ross instructs all who have anything to do with the dying to touch, touch, touch their loved ones as they leave this world.

If touching the dying is an altruistic act, touching the already dead may not be quite so kindly, and a bit more self-serving. Until recently, in rural areas of England and America, the belief persisted that touching the corpse in a coffin would dispel evil spirits and assure good fortune to the living toucher. Before the coffin was closed, the family would invite special friends to touch the body, and even passing strangers were sometimes offered this shot at good luck. No one would think of refusing the dubious honor. And, if you happened to miss the corpse, perhaps you could arrange to get to carry the coffin to the cemetery; if so, and if you had warts, you'd be in luck. Nothing as good for warts as touching coffin wood (or so went the tradition). If you'd be allowed to touch the corpse *and* the coffin, you'd be virtually promised a long, wart-free life.

The black community, with great intelligence, rarely questions touch as a survival technique. In complicated modern society, touch often tends to become polluted with too much portent, too much power, too much purpose. In the Afro-American tradition, however, touch still tends to remain delightfully simple and expressive as primarily an elemental means of connection. While their neighbors join therapy groups to get in touch with their psyches, many very sophisticated young black people simply go to church, as their ancestors did, and derive pleasure from joining hands as they sing hymns, says Dr. James P. Comer, a professor of child psychiatry at Yale.

Alex Haley, author of *Roots*, speaks of his grandmother, Cynthia Palmer. "Part of my collective memories of her was the way she was always touching you or petting you or straightening your shirt," he reminisces. "I remember the sensation of not being in her presence for more than a minute or so before she'd find some reason to make contact with you. To me, this was a reassurance of her caring, love and concern for your well being. Even when she was grumbling something like, 'Boy, you've gotten all dirty,' her fingers would still be moving lightly and deliberately as if she was saying with them, 'I care for you, I love you.'"

The palm-slap-and-slide, which originated in the black culture, has been copied by much of America's youth, who use it as it was meant to be used and that is a way to say, "Hey, man, we're great, we understand each other, catch you later." The "solidarity" handshake, which also originated in the black community, says the same thing. Touch is so basic to the black community that the only time many black men in modern societies hesitate to touch is when a tender, delicate, loving hug is called for. Macho palm slaps are not open to misinterpretation. Tender hugs are.

Tender hugs are not a problem for many modern Israeli youths. Jonathan Half is a tall, blond nineteen-year-old soldier in the Israeli Army, always on call, always training, always on the ready to protect from enemies. The esprit de corps that exists among men who have grown up expecting attack from unexpected quarters is wonderful to see and feel.

Jonathan says, "On Shabbat, we dance, our men, just as our grand-fathers did, linking arms, holding hands, wildly, joyfully! The children raised in kibbutzim are subject to hugs and touching from every member on the kibbutz—not only their parents. Everyone is their parent and everyone touches. We embrace our fathers as easily as we do our mothers, our friends as quickly as our cousins. We spend our adolescence preparing for defense and many of us don't have time for great intellectual ponderings. We always have time, though, for rough-housing, dancing or hugging. To be Jewish is to be a toucher; we touch each other as we breathe."

Science has begun to tell us what many cultures have known all along. Some neurophysiologists, like Dr. James A. Prescott of the National Institutes of Health, suggest that societies that don't touch may create aggressive people and cause violent behavior. A human being missing out on nurturing body pleasure may seek substitutes like pornography or physical violence, and if the touch deprivation is severe, structural damage may even occur in the dendritic branches of the brain cells themselves.

There are many who agree with Prescott.

Allen Forbes, an anthropologist at Brown University, has observed loving as well as aggressive touching all over the world. "Too little loving and too much aggressiveness is what makes wars," he says.

"Many people are innately aggressive because of an inability to touch and be touched. Their societies frown on touching, especially within their own sex. And so they find ways, if they are reasonably healthy, to touch in an acceptable manner. The sports arena is often little more than a sublimation of touching."

Take butt patting. If a burly, macho football coach were to pat the butt of a twenty-two-year-old running back when both were at a cocktail party, he might be looked at with amazement by all present. But, on the football field, a civilized kind of battle takes place and ritual provides for the modern/ancient giving of support through touch. As each star player lopes onto the sporting field at a crucial juncture, the coach, emulating his ancestors' ritual, pats his husky player's butt. It's an absolutely imperative move and both player and coach would feel oddly bereft without that loving butt pat. And the touching ritual is extended even further into the game by the "high five" ceremony, which millions of Americans watch every time they sit glued to their television sets during football season.

His butt patted, the star goes out to score a touchdown and, when he does, in his moment of victory, he comes running back to his group, where he jumps in the air, his right hand extended on high. The other nearby athletes also jump, and with their own arms extended high, touch the victor's raised hand. High Five. Forbes notes that sports touching might well derive from wild dogs or wolves on the hunt. Like modern hunting dogs, before their morning quest, they nuzzle, nose, and rub bodies, "have a little huddle," says the anthropologist. This touching ritual is like a consciousness-raising encounter. When the dogs have made their kill and they've scored—like the football players—they return to the pack for another nuzzle. This kind of touching for dogs, wolves, and men, notes Forbes, is meant to say, "Hey, we're wonderful!"

"I've even seen horses nuzzling, particularly when they're under stress," says Forbes. "And in the face of a death in the community, I've seen Australian aborigines come together and form a solid mass of touching. It's an absolutely fundamental act, part of the human and animal baggage for over a million years. We simply cannot survive without touch."

Nevertheless, even to a casual observer, it would be clear that touch means different things to different people. If Balinese adolescents often fall asleep just leaning on each other in a crowded room, New York subway riders don't. It's worth your life to fall asleep while leaning on a New York subway rider. Two researchers, Argyle and Dean, published a paper in 1965 that put forth their "equilibrium" theory of interpersonal distance. Once people have determined what a comfortable social distance is, they'll maintain it, holding a drink or a cigarette perhaps to ward off incursions of touching, leaning back if the person to whom they're talking comes uncomfortably close. In different societies, a comfortable social distance means disparate things. At a UN party, the Mexican ambassador might reach out to touch the English ambassador, who steps back, feeling violated, right into the path of the Italian ambassador, who has just good-naturedly cuffed the German ambassador on the shoulder, much to the latter's dislike.

But it's okay to fall asleep while leaning on a Parisian in a Métro, where, Ashley Montagu reports, "Passengers will lean and press against others, if not with complete abandon, at least without feeling the necessity either to ignore or apologize to the other against whom they are leaning or pressing. Often, the leaning or lurching will give rise to good-natured laughter and joking and there will be no attempt to avoid looking at the other passengers."

Traditionally, those in every society from an upper class have more leeway to touch anyone they like and those from the lower classes may touch only each other. In India, this concept was carried to its most objectionable level by the concept of the "untouchables"—those members of the lowest Hindu caste who were held to defile a member of a higher caste on contact.

In America, even though our sense of "class"—as in English peerage—is hazy, it's still considered vulgar to touch strangers. Anyone who inadvertently touches another in a crowded elevator or while passing on a street (or anywhere it's likely that strangers might make physical contact) must offer an immediate apology or expect a dirty look. It's almost as if the unwillingly touched person has been defiled.

An oversensitivity to physical contact is most often influenced by culture and the immediate family, but researchers Lee and Charlton point out studies that have shown there may be something else—even biological—involved. Prison inmates with histories of violent crimes, when tested on the comfortable-personal-distance theory, have felt that the interviewer was "up in their faces" and "crowding them" at distances most people considered reasonable and unthreatening. Lee and Charlton speculate on whether this feeling of needing extra room was due to their problems with the law or indeed a precipitating factor of the problem.

It hardly takes an anthropologist to see that people of different cultures feel different about touching. Walk down the street in Rome

and watch the young, virile men linking arms as they strut their man-
hood and joy in their friendships. In another city, they'd be labeled
homosexual before they had a chance to flex a muscle. Watch young
Frenchwomen, arms linked, sides pressed close together as they stroll
down the Champs Élysées. Wait. Watch that Frenchwoman who has
just linked arms with her American visitor. The American girl is distinctly uncomfortable; all that touching makes her feel peculiar. When
she was a little girl in Brooklyn, she'd walk with her playmates with
their arms slung together in easy camaraderie, but the moment she hit
sixteen, she almost never touched her best friends. You held hands with
boyfriends, not girlfriends.

Because the world has grown smaller, intercultural touching is now
more possible but often misunderstood. While a Venetian teenager
kisses his best buddy with gusto and warmly receives affection, he'd
better think twice before he kisses his visiting American pen pal in the
same manner. An American businessman recently ran into this form of
culture shock. He'd been invited to the home of a large Hong Kong
pocketbook manufacturer—his Chinese counterpart. The invitation
was a big score and he hoped to set up a trade agreement. The dinner
party was gracious and replete with good feeling. Returning home, the
American spent the better part of a year trying to reach his new business colleague on the telephone. No dice. Finally, reconstructing the
entire party, step by step, with an expert in things Oriental, he pin-
pointed the problem. Leaving the party, he'd planted a kiss on the
cheek of his host's wife to try to express, with American warmth, his
thanks. American warmth failed. Touching was a breach of etiquette,
and the unwanted touch had been humiliating, especially in the presence of others, to both his host and the host's wife. The partnership was
dead before it started.

The late Professor Sidney Jourard, a psychologist at the University
of Florida, was struck by the touching differences he'd observed in
multinational students, and one year Dr. Jourard made a trip around
the world. His purpose was not to see the ancient ruins of Rome or the
fabled mosques of Istanbul, but to count touches. In hundreds of coffee-
shops in hundreds of cities, Dr. Jourard watched every time he saw a
twosome sharing a coffee. During the course of an hour, he jotted down
the number of times one person would touch another. The results?
London failed miserably with not one touch between people. Gaines-
ville, Florida, didn't do much better, with only two touches to its credit.
Paris? Ah, Paree. In Paris, a twosome touched 110 times. And in San
Juan, Puerto Rico, two coffee drinkers touched more than 180 times!

Perhaps Jourard's experiment wasn't the most closely controlled scientific study of its genre, but the results were astoundingly accurate. There's little question that societies like Italy, Mexico, Spain, Egypt, Russia, Turkey, Greece, and France are contact societies, say sociologists. The Dutch, Germans, Japanese, English, and Americans are not touch-loving as a group—at least, not in public.

But Egypt, Turkey, and many other Moslem countries also have strong touch taboos. Although touch may be practiced quite openly between members of the same sex—and certainly with children, who are always jumping in laps to be hugged—public contact between the sexes is strictly frowned on, even in countries where veils have long since been abolished. And if you should happen to be a veiled woman in a strictly orthodox Moslem country—say Saudi Arabia or Iran—woe be to the adult male who even dares to look at you.

Touch carries power—such great power that its rules in almost every society are apt to be as circumscribed as those for sexual intercourse. In some communities, touch is only an invitation to bed or to battle; in others, it is used as a fond gesture of friendship or kinship as well as for mutual sexual interplay. It can be used as a power play that separates or it can be used to strengthen the bonds of affection and love.

Through the ages in almost every culture, the power of touch has also been used to heal. Eons of experience have produced extraordinary healing systems from many cultures—some of which I will describe in the chapters to come.

Touch may leave its most profound imprint in the daily, intimate relationships between mother and baby, parents and children, husbands and wives and, of course, between lovers. Indeed, as our Western civilized scientific cultures are beginning to rediscover, these ordinary, day-to-day, loving touches are more vital to our healthy growth and development than we ever imagined.

CHAPTER TWO

LOVING TOUCHES

BIRTHTOUCH

. . . we feel the tenderest touch.

JOHN DRYDEN

Her belly is out to *there*. She stands, lost in thought, waiting for a bus, and without even being aware of it, she rubs and rubs and rubs that baby in her belly. Her hand smooths the clothes that stretch over the rounded shape of the fetus that lies within. She gently fingers the place where a tiny foot has kicked. She caresses her rounded front with infinite delicacy, as if to soothe her baby.

The scene is repeated thousands of times a day. Women who carry unborn babies simply can't stop touching them—even though the baby is buried under layers of clothing, skin, fat, and fluid.

"It's almost as if they're communicating with them," says Elisabeth Bing, author and famed educator in the art of childbirth. "There's no doubt touch calms the fetus and mothers-to-be seem to instinctively know it. There is an electrical field, I'm quite convinced, that is passed from hand to hand, body part to body part. When a father assists in his child's birth, for instance, and rubs the mother, it eases the birth process and provides a wonderful communication between the parents. Touching helps one part of the body relax while another part works. And touching is a pain reliever. Think—what do you do when you have a headache or a backache: you rub the part that hurts. It's no accident that being touched soothes the unborn fetus as well as its parents."

During birth itself, the mother's laboring action touches, pushes, pulls, and hugs her baby throughout its trip down the uterine canal.

Some researchers have gone so far as to suggest that without pre-birth contractions—as when a child must be delivered by Caesarean section—there are greater risks of developing physical and emotional problems. Dr. Thomas Verny, an obstetrician, has written that the massaging the baby receives as it moves through the birth canal is the baby's first encounter with sensuality. A baby deprived of this introductory massage often has "trouble with the concept of space. He does not seem to know where he begins or ends physically, so he is prone to be clumsy [and he] demands, indeed requires, continual stroking and hugging." The doctor says that what he calls "Caesarean disorientation" is a sign that touch gives us a sense of our separate selves as well as a sense of someone being there for us—the twin bases of mental health. All is far from lost for Caesarean-born infants, however, and other experts offer ways to compensate for the missing uterine stimulation. Dr. Richard Hausknecht, gynecologist and obstetrician at Mt. Sinai Medical Center in New York, and Joan Rattner Heilman, in their book, *Having a Caesarean Baby,* suggest that plenty of mother-baby contact immediately following birth seems to "touch away" the "Caesarean disorientation." Most Caesarean babies are taken to the hospital nursery immediately after birth and are kept there for twenty-four hours before the mothers see them. "The babies are treated like surgical patients—in other words, invalids," says Dr. Hausknecht. Bad news. If there are no medical problems, advises Hausknecht, parents can bypass this practice by telling the doctor, "before, during and after birth," that they wish to touch their babies immediately. Not only in the first afterbirth hour but in extended periods after that, touch can nourish the systems of both Caesarean and vaginally delivered babies.

Maddy Haut started to feel her first labor pains during the night of April 17, 1986. Jeff drove her to the hospital to begin what was to be almost a seventeen-hour labor. When they arrived at 3:15 A.M., Maddy climbed up on the examining table and her water broke. From then on, it was touch, touch, touch throughout a long and sometimes frightening evening. Jeff and Maddy had taken the popular Fernand Lamaze course, in which both prospective parents learned controlled breathing techniques, which help a laboring woman respond actively to her contractions. Husbands learned how to help by breathing in rhythm and rubbing their laboring wives' necks, backs, ankles; how to mop a damp forehead; how to nurture with touch during this most extraordinary experience both would share.

Jeff's hands were hardly ever away from an increasingly nervous Maddy as her obstetrician gently talked about the possibility of a Caesarean birth. Finally, though, the cervix dilated sufficiently for a vaginal delivery. The doctor left to prepare an epidural injection which would be used for anesthesia.

"I was terrified of the injection," said Maddy. "I told the labor nurse that I had passed out having my ears pierced and she understood immediately. She held my hand hard with both of hers and the shot didn't hurt at all. I never felt it.

"During the delivery, Jeff and the nurse mopped my brow, massaged my shoulders—Jeff kissed my face. When it came time for the final push—and I was exhausted because it was 6:45 in the evening by then—the nurse held my entire body as she would a child's, and I pushed and I *know* that, if it weren't for her, I never would have been able to push my Jamie out.

"Afterwards, I was shaking. The injection, the effort, the emotion —I couldn't stop. Jeff had left the delivery room to tell the good news to everyone and the nurse, bless her, held my baby on my belly, just kept Jamie there on me so that we could feel each other. I'll never forget that nurse."

Dr. David Chapin is an obstetrician and gynecologist at the Brigham and Women's Hospital and the Massachusetts General Hospital in Boston, Massachusetts.

"It is clear to anyone who has ever been involved in prepared childbirth that hands on the laboring woman is of enormous value in making her labor more tolerable and in reducing her need for narcotics and anesthesia. More and more of us are realizing that the doctor's hands on his patients are part of his therapeutic tool kit. The husband's hands are by far the most meaningful but a friendly nurse or doctor will do just fine."

There is little disagreement among today's obstetricians that, whenever feasible, fathers should be included in the birth experience. Not only can he provide enormous help to his wife but, as a father, he experiences that unforgettable moment of touching his own creation as he or she draws the first breath. A strong bonding begins immediately.

However, such obstetrical practices are fairly new to the industrialized, "scientific" West—though variations of birthing touch have been used in other cultures and primitive tribes since time began. Only a generation or two ago, it would have been considered unsanitary and unsavory to have an observing male present during such a womanly process. Often women were left for long hours of labor in agonizing isolation or together with other groaning mothers-to-be. The first stage of labor was painful and terrifying; the second stage was often obliterated entirely through anesthesia. One of the greatest moments in life—for mother, father, and child—was bypassed.

We now know that birth touch—contact before, during, and right after birth—is an affirmation of a loving world for all involved. Parents touching a newborn and each other set the stage for a child's lifetime sensitivity to tender touch.

For information on finding classes for natural childbirth, see "Resources," page 189.

TOUCH AND BABIES

Parents of newborns might stroke and caress their infants from some atavistic knowledge that touching feels good to all, but the benefits of such stimulation transcend mere emotional bonding. Psychologist Tiffany Field headed a revealing study at the University of Miami Medical School which was reported in *Science News* and that study has ringing results! Twenty randomly selected premature infants were given touch and movement stimulation for three fifteen-minute periods a day for ten consecutive days. The babies that received the touch stimulation ended up gaining 47 percent more weight per day than a control group of untreated infants even though both groups received the same amount of calories in food intake. The touched babies were also more awake and physically active, could better orient themselves to new stimulus, were less likely to cry one minute and fall asleep the next and were better able to calm and console themselves.

Coincidence? Unlikely. An eight-month follow-up showed the touched babies still thriving better. They were longer, heavier, and had fewer neurological problems than the unstimulated infants. What kind of power does this thing called touch possess?

A very potent power. If you're talking about the magical power of touching, I have to tell Ashley Montagu's story about Old Anna.

Not long ago in this country institutions flourished that were known, unromantically enough, as foundling hospitals. They were an unmitigated disaster and no one could understand why because standards of cleanliness were high. Still, reports Montagu, as late as the second decade of the twentieth century almost every single baby under a year old in an American foundling hospital died. Montagu further says that in an investigation of children's hospitals in 1915, the noted pediatrician Dr. Henry Dwight Chapin reported that in all but one institution, in ten different cities, every single baby under two had died.

All this just happened to coincide with the "Don't Touch" era, the years of scientific sterility. Touch was the shunned sense. Babies get spoiled if you touch them when they cry—well, don't they? The hospitals were very civilized, very modern, very nontouching. In the nurseries, babies were fed the very latest in infant formulas. Everything was clean as a whistle; you could eat off the floors. Germs were not allowed, that's all. Hands off the newborns.

But the babies were dying, melting away as though in the burning heart of an epidemic.

Then, Dr. Fritz Talbot went to Germany. He visited the Children's Clinic in Düsseldorf and made rounds with the rest of the doctors. They had nothing we didn't have, he was pleased to see. He'd have a good report of German clinics to bring home.

Something odd caught Talbot's eye before he left.

It was a very heavy old woman. She wandered the wards every day, and every day she had a scrawny baby on her arthritic hip. She nurtured, she cuddled, she stroked the babe. Who was she? A lost grandmother? And why was she handling that child so much? Didn't she know the rules of sterility and overspoiling?

"Who's that?" asked Dr. Talbot.

"That," sighed Dr. Arthur Schlossmann, the director of the clinic, "is Old Anna. When the medicines have all been tried, and we've done everything that science has available, and the baby is still sick, we turn it over to Old Anna. And she cures it."

Too simple. Who would believe it? Touch therapy was a barbaric theory, contradicting the new, brilliant voices of modern medicine. Still, Talbot came home convinced that TLC—tender loving care—was better than clean hands. He had a hard row to hoe.

Gradually, other pediatricians began to look into the remote possibility that there just might be another approach. Investigations were in order.

What the investigations turned up was mind-boggling. Babies, even from the most indigent homes, babies who were experiencing squalor, unsterile homes, poor and insufficient food thrived—*if* they had parents who held them, and held them plenty—shades of Old Anna.

When the doctors examined their own hospitalized, carefully controlled infants, they saw listlessness, lack of appetite, yellow-gray pallors, lack of eye fluids, faint heartbeats, and shallow breathing. They saw clean but sick kids. What's more, the sicker the kids got, the less they'd respond to any touching at all.

The "dying" epidemic wasn't only affecting foundlings, who had no parents to cherish them. Much-loved babies who had to be hospitalized for infections were involved in the terrifying attrition. One of the reasons there were so many babies hospitalized in the first place had to do with their basic food—milk. In the early 1900s, milk was sold over the grocery counters from dippered, unrefrigerated five-gallon cans. Diarrhea and respiratory infections ran rampant among the youngest population, and many were sent to the hospital by frantic doctors and parents. In the interest of cleanliness, isolation wards were born. Germs were the enemy, and parents and friends harbored them; therefore, visiting hours in most major children's hospitals were limited to less than an hour a week.

Children were sick and scared and alone, and untouched except when absolutely necessary. Huge barriers separated them from each other. In the maternity hospitals, the newborns were kept isolated.

During this time, it was a tossup which would kill a hospitalized child first—germs or touch deprivation. No one yet knew that you had to, quite literally, be touched or die.

Bellevue Hospital in New York, to its everlasting credit, was one of the first to throw in the towel. All that perfection had to go. A new routine was instituted. Every hospitalized baby would be picked up, held, and mothered every day, several times a day. Parents would be encouraged to come in and hold their own children. Foundlings would be held by nurses and volunteers—like Old Anna.

By 1938, mortality rates for infants had dropped to less than 10 percent of earlier, tragic figures. Old Anna should have won a Nobel Prize.

The doctors discovered that it was starvation that was dropping the babies like flies—not food but touch starvation. They gave the disease a name: *marasmus,* from the Greek "wasting away." A vital nutrient in the human diet, touch, keeps people from wasting away.

All mammals except man touch-prod their babies into sentience immediately after birth by licking, nuzzling, pushing, and pulling. What the mammals are actually accomplishing is a simple stimulation of their babies' respiratory and digestive systems. The prodding encourages everything to work. Without such touching, their young would die. Until recently, it was felt that this early touching in mammals was merely an atavistic grooming instinct, but today we know it's much more.

When an infant is born, its field of vision is narrow and blurry. Although hearing is fully formed, it takes almost a year to "place" sound so the baby can rely on it for learning. Smell and taste are also fully developed, but very unsophisticated. Time and experience will be required to make sense of their complicated messages.

But touch . . . touch is all there, utterly accurate, simple, clear-cut. It is the newborn's first telephone to the world. If the first touch messages are brusque, the baby gets a too-busy-for-you signal. If the first touch messages are loving, the baby gets a hello-we're-glad-you're-here-get-growing signal. If the newborn is touched too roughly or too seldom, his information about the world will be faulty and confused. The lack of touch will soon affect his mental and physical stability.

The infant who is touched often and joyfully has the world on a string before he even gets his first Dick and Jane book.

The way the infant touches out to the world also defines his future. The world unfolds either terrors or challenges, depending on what the baby's touch tells him. If he touches something unpleasant, he begins immediately to process this information for future and forever reference. This is, indeed, how babies find out that hot is not great and teddy bears and soap suds are. No baby is satisfied with just looking at an object; he has to touch it, put it in his mouth, nuzzle it, sit on it. Millions of different nerve endings in his skin pick up touch messages and transfer them to the brain. The nerve endings inform him about the boundaries of his body—where he ends and other things begin—as he touches. The nerve endings allow him to sense the world—his mother, his crib, his bath—as he touches. He uses his mouth, which has a zillion nerve endings, as he touches objects to this part of himself.

As babies, we learn to identify objects by touching them, but we also learn to differentiate between people by the way they touch us. If your baby had language, you'd be amazed to see how he could touch and instantly recognize almost everyone in his circle. Dr. James Gibson of Cornell University demonstrated that people can detect movements across the skin as slow as a millimeter a second (the same speed as the minute hand on a clock). The skin senses the location of distances on its surface more accurately than the ear can read sound distances. The infant reads his surroundings and his people by the way they feel on his skin. His skin becomes his education, in short. As little as a baby is, a piece of his skin the size of his mother's eye holds more than 3 million cells, 100 sweat glands, 50 nerve endings, and 3 feet of blood vessels. He is, as they say, *equipped!*

Sometimes, the power of this equipment is shockingly dramatic.

One special crowd of the "littlest lovers" are really the very littlest, and touch, in the most urgent sense, saves their lives. They're heart-breakingly tiny, this premature set, and their survival is a day-to-day struggle.

In the newborn "preemie" room, in an intensive care unit of Massachusetts General Hospital, tiny babies lie trembling under probing lights. They need every ounce of their shivery strength to survive in a tough world for which they're not yet prepared. Tubes, patches, and wires cover their bodies, which are sometimes no bigger than a father's hand. Every now and then, a preemie becomes agitated—something throws him off, either a harsh sound, a light, a draft, an interior imbalance. He lies on his back in the respirator, tiny arms reaching out, fragile neck tensing as he desperately struggles to regain his equilibrium. He is disorganized, unbalanced in a world that suddenly lacks the support of the womb, which by all rights he should still be enjoying. The buzzer connected to his electrocardiogram goes off, his oxygen level drops precipitously, and something must be done immediately. The staff jumps—it's that touch and go. A nurse comes running. She places her finger on a baby's least-wired, least-patched, least-tubed section of skin, and she rubs it very softly. Often, that's all the baby needs. The warning buzzers quiet. The baby relaxes and is able to concentrate on living.

Dr. David Todres, the director of the newborn and pediatric-care unit at Massachusetts General Hospital, says, "Touch is seen as a need rather than an indulgence and the need is shared among the infants and the parents, the nurses and doctors who must bond with them if the babies are to survive."

Dr. Heidelise Als of the Harvard Medical School, a research psychologist at the Child Development Unit of the Boston Children's Hospital, explains further: "The 'preemie,' who comes out of the womb too early, has suddenly been thrust into an environment that does not have the earlier containment that the uterine sac, the fluid and the surrounding skin provided. All of this prenatal touching helps the fetus organize his motor system. Abruptly, it's gone. In the incubator, lying on a flat surface, he's getting very minimal tactile contact."

The preemie's body posture feels awkward to himself and it looks awkward to others, says the doctor. He may try to get himself in a more "flexed and inhibited" position but, in doing so, he becomes exhausted and compromises his whole system.

It's very important to help these infants "organize," says Dr. Als. "One way of doing it is to touch the infant on his open, splayed palm; he begins to put his fingers around your finger and as he does this, he brings his arms in, relaxes his shoulders and neck and finds a more comfortable position. This alleviates a very dangerous strain on his breathing and on his heart."

Another way of helping a preemie to gain a "grounding" is to touch the soles of his feet or to draw a line with your finger down the center of the front of his body. "When the babies are brought back into equilibrium," says Dr. Als, "you may gradually withdraw the touching supports and they are often able to sustain the balance for long periods of time —or until, at least, the next unsettling event, like a light suddenly flashed on in the nursery."

The evolutionary aspect of grounding an infant is of particular interest to Dr. Als, who says that all primate newborns are carried with "their bellies towards their mothers' bodies"—a total touching of the infants' frontal surfaces, as the babies cling with hands, feet, and mouths. One of the reasons for this, speculates the doctor, may be that the temperature-regulation cells of all primates, including man, are more efficient on backs. Warming the newborns' fronts "presumably makes up for the lesser warmth that's experienced on the front of the body."

Many hospitals, finally knowledgeable in the efficacy of touch, now arrange for premature or very tiny babies to sleep, cuddled back and front, in a sheepskin that stimulates the skin and tends to mold the infants into a simulated mother-touch security.

In a study of children suffering from apnea (suspension of respiration), rubbing the infants' extremities produced a significant decrease in the frequency of the apnea, during and after the stimulation. Babies placed on gently oscillating water beds also fared better with apnea.

In addition to restoring equilibrium, touch can make a premature infant more active and freer of pain and help him grow, says Dr. Als.

In Cambridge, England, researchers found that preemies who slept on sheepskin gained half an ounce more each day than those who lay on ordinary cotton sheets. At Duke University Medical School, Dr. Saul Schanberg and his colleagues have also demonstrated that touch affects growth significantly. "It's dramatic," he says, "the handling of preemies improves their weight gain markedly."

Schanberg started out with rats. Studying hundreds of them, he found that touch triggers elaborate biochemical processes. Like all other mammals, a mother rat strokes her babies with her tongue when they are born. This releases growth hormone and activates the enzyme ornithine decarboxylase (ODC) in the pups' brains and other organs. As a regulator of protein synthesis, ODC is necessary for growth. Schanberg then separated the rat pups from their mothers for short times and what happened? Infant rats refused to grow. Their ODC levels and their growth-hormone levels fell, not from lack of nutrition but lack of touch. Then the researchers separated another group of rats from their mothers, only this time they stroked the babies with tiny paint brushes. The biochemical drop was reversed.

"Active coddling is the key," says Schanberg, who adds that pups deprived of a mother's TLC do not grow even when growth hormone is injected.

Schanberg turned his attention to human babies. For ten days, he and his colleagues gave daily massages to premature infants in intensive care. Compared to non-massaged infants, the touched babies gained 47 percent more weight each day, showed more mature infant behavior, appeared more alert and active, and were discharged from the hospital an average of six days earlier.

Biochemically, Schanberg believes, the baby rats and the baby human children are very similar. Studies have been made of a condition known as psychosocial dwarfism; in this condition, deprived children are stunted in their growth and don't respond to growth hormone, as in the tests with the baby rats. If put in a loving, positive environment, with plenty of touching, they sprout inches. Schanberg speculates that babies of every genre who aren't touched and who sense their mothers' absence, appear to shift their energy from growth to simple body maintenance. They just maintain the status quo—no more. Babies who don't experience early touch, believes the researcher, somehow purposely stop growing—as difficult as that may seem.

What about children who are born with serious birth defects? Or people of any age who are injured? Touch seems to offer therapy there, as well. For example, Dr. Emanuel Chusid, Medical Director of the Mental Retardation Institute of the Westchester County Medical Center in New York, reports that many Down's syndrome children will walk earlier if they experience a lot of maternal hugging, touching, and stroking. And doctors at the Harvard Medical School report that decreased spasms and more normal muscle action can be achieved in some cerebral-palsied people through a "brush-touch" sensory-stimulation technique. At the Shriners' Burn Institute in Massachusetts, nurses reach inside the clear plastic flap of the isolation tents where badly burned patients are placed to avoid further contamination of their burns from the air. The nurses wear long gauntlets and they handle the patients wherever they're not burned, rub their hair, stroke a finger, anything, to "let them know we're here." Touch, which is a "fundamental concept of nursing," says nurse Anita Wright, "actually heals. There is a tremendous relevancy between human touch and a relaxation of pain."

But infant touch and massage are not just for preemies and children with serious defects. It is golden for all normal babies as well—and for their parents!

Amelia Auckett, a nurse in charge of the Infant Welfare Center in Frankston, Australia, says a gentle stroking of a baby's lush skin has been found to have immeasurable benefits for both baby and massager—whether the latter be a mother, nurse, father, sibling, or loving pal. She has written a wonderful book called *Baby Massage,* and in it she explains how the contact creates a "metaphysical energy flow" between mother and baby. Give or take the metaphysical energy flow, the colicky, stay-up-all-night baby adores massages because it makes his stomach feel better. The poor teether, the constipated, wretched infant, the baby who doesn't find it easy to nurse or to be weaned—all these benefit from the sweetest contact between pliant, infant flesh and loving adult fingers.

All this touch may be news to us who live in some section of the industrialized world, but it seems to be old hat in many other cultures.

In Senegal, for instance, the midwife massages the vaginal opening in the mother's body through which an infant must pass, and then the scalp of the almost-born fetus to ensure a slippery, easy passage. Only a week after birth, the babe is massaged, rather roughly by American standards, by the whole tribe, who take turns manipulating flesh, pulling and pushing baby limbs. The Senegalese baby is afterward in almost continuous contact with its mother, and studies have shown it to be more precocious and coordinated, as it grows, than its European counterpart.

Amelia Auckett reports that in Nigeria babies are massaged frequently from birth until one year of age. In Malaysia, mothers are routinely massaged on the second or third day after giving birth, then every day for six weeks by their mothers, mothers-in-law, or even grandmothers. Massage dilates these new mothers' blood vessels, which improves general circulation, which in turn does great stuff for tension, muscle spasms, and congestion—not to mention fatigue. It has been noted by anthropologists that Malaysian mothers return to their full strength and vigor remarkably soon after birthing. In New Guinea and New Zealand, among other nations, most babies are touched, petted, massaged from the moment they see light, and, say many anthropologists, these babies have been found to be alert, happy, calm, cramp-free and even, think some, particularly intelligent. In Uganda, where it has been said that babies are extraordinarily advanced and intelligent, they are massaged, fondled, and sung to from birth. Indian babies are regularly massaged in the winter with diluted mustard oil and in the summer with coconut oil. Fijian, Balinese, and Venezuelan babies are massaged and touched and rarely leave the warmth of their mothers' skin surfaces.

From our own Western scientific neo-natal studies—most of which have been done within the last twenty years—we have discovered that early touch not only helps an infant develop physically but is the key in developing a special relationship between parent and child called bonding.

Drs. John H. Kennell and Marshall H. Klaus, the two researchers from Case Western Reserve Medical School who have spurred the highly influential bonding movement in this country, define "bonding" as that *unique* attachment that begins between baby and parents in the "sensitive period," sometimes within the first few moments or hours after birth, but surely within the first seven days. Touch in this period creates a close, rich relationship that can endure through time. In fact, say Kennell and Klaus, "This original parent-infant tie is the major source for all the infant's subsequent attachments and is the formative relationship in the course of which the child develops a sense of himself. Throughout his lifetime, the strength and character of this attachment will influence the quality of all future ties to other individuals."

Pretty heady stuff. If it's true, one certainly wouldn't want one's babe whisked away from the delivery room to the hospital nursery to be touched another day. Klaus and Kennell deplore the "another day" theory. "Fondling, kissing, cuddling and prolonged gazing," the tools of early touch bonding, help parents fall in love with their infants, an act that's not nearly as instinctive as one might suppose. And touch now is better than touch later. The first week is an enormously important touch time.

In their research Klaus and Kennell found that this early touching even influences intellectual development! The doctors matched control groups of children whose mothers had given them a great deal of cuddling and holding within twelve hours of birth with groups of children who did not receive this early touch contact. The families of the children were similar in economic status, education, background, everything except the touching. When followed up at regular intervals in their lives, the children in the early-touch-contact group had higher reading-readiness scores and even significantly higher I.Q.s.

"Their very language development was markedly different," says Dr. Kennell. Following two similar groups of babies and mothers with the only significant difference being that of extra contact in one group, the babies who were touched more frequently ended up with richer speech patterns. The less frequently touched babies, as they developed speech, used word arrangements you'd see in a telegram: "See bird." The more frequently touched babies might say, "See the pretty red bird with the black wing."

Dr. Kennell observes further that "terribly sick babies who have had lung diseases or abdominal surgery are not touched as much as well babies in a hospital because of all the tubes and such." Quite a few of these sick babies seemed, at three and six months, actually retarded. Weight gain was poor. Worse yet, the babies' own mothers were often so fearful that their sick babies would die they psychologically "gave up" on the infants, and touched them as little as possible.

"But," says the doctor, "if we provide someone who will act like a mother, fall in love with a sick baby, do a great deal of stroking, touching and holding whenever medically possible, a great many of these young-sters will go from seeming retarded to being responsive, smiling, weight-gaining babies."

All these professional neo-natal studies have captured the hearts and minds of eighties parents—many of whom now take courses in infant massage from special infant massage trainers. "Ashley Montagu probably set the stage for all this," says Ramelle Pulitzer, a pregnancy and postpartum exercise specialist connected with Santa Monica Hospital in Los Angeles, "but then came Frederick Le Boyer with his great book, *Loving Hands,* and the work of Marshall Klaus which showed us how important it all was. Klaus's latest book, *The Amazing Newborn,* is incredible. I would no longer think of giving a post-partum exercise class without including the babies and showing the mothers how to do infant massage."

Neither, I guess, would Jane Fonda. In her pregnancy and postpartum exercise book and videotape, her pregnancy-exercise specialist—Semi Delyser—includes detailed instructions in the art of baby massage.

"It's beginning to move very fast throughout the country," says Pulitzer, "and I, for one, would like to see it taught to every new mother. It allows the baby and mother to have a relaxing dialogue. It's very intimate but it doesn't impose. Through doing it, a mother can really learn to 'read' her baby or young child. Rather than teaching her games or showing her flash cards, you do a mutual thing and learn about each other. You have a chance to talk to each other through touch."

"Baby Fair"—a new traveling show on pregnancy and early child-hood—also seems to be spreading the word—and fast. Its two most popular presentations are on sleeping problems and infant stimulation—both of which include infant massage. "Baby Fair" will shortly make nine presentations in major U.S. cities.

"And," says the enthusiastic Pulitzer, "practicing the massage with my own young children—three and five years old—has reduced the noise level in my house incredibly. Why? Because you know your children so much better and it teaches you how to talk to them—not five or ten feet away—but within arm's reach and at their own eye level. It's amazing the difference the knowledge you get from the massage makes in your day-to-day life."

One of the best times for regular infant massage is after the bath. The stroking should be relaxed and without much pressure—just a gentle, continuous, rhythmic moving of the hands over the body and limbs. But, if a baby can't get to sleep or seems to have trouble digesting, that's a good time, too. "Little babies," says Pulitzer, "are balls of pure sensations so they're stressed out at the end of the day. Nothing helps so much as a gentle massage."

Laurie Evans, another instructor-trainer in the art of baby massage, says that touch creates a bond between father and child as well as mother and child. It is extraordinarily effective, says Evans, on babies who have asthma or other respiratory or circulatory problems. And most practitioners agree that the bonding process is served better by massage than just casual hugging or patting. The massage seems to "ground the infants," who feel better about their bodies, more at ease in their skin and the new world into which they've been plunged.

Can there be such a thing as too much touching of infants? Can too much cuddling make a baby dependent or somehow spoil him?

Apparently not. According to many experts like Mary D. Salter Ainsworth, Ph.D., a professor of psychology at the University of Virginia in Charlottesville, "It's nearly impossible to give a baby too much close contact in the first few months." After she conducted a year-long study of twenty-six infants and their mothers, she discovered that babies who were touched the most cried the least—even when left alone. Apparently cuddles point the way to independence.

And good skin. Ashley Montagu has discussed some telling connections between the skin and emotional problems. People who experience sexual disorders frequently experience skin problems, and the common link may be an early lack of cuddling and touch.

In *Pediatrics*, Dr. Maurice J. Rosenthal reports the results of his research, which tested whether eczema might occur in predisposed infants because they failed to receive enough cuddling from their mothers. He investigated twenty-five mothers with children under two who also had eczema and determined that, indeed, the children had been touch-deprived children. It's almost as if, said one expert, the skin was "writing the message of its early Deprivation on its own surface."

Stress is written on the skin's surface more frequently than we'd like to acknowledge. Outbreaks of pimples or eczema, sweaty palms and upper lips, and blushing are all messages of the skin, which can frequently reveal problems more accurately than an hour on the psychoanalytic couch.

How we touch our children is preserved in their memories forever —their skin memories as well as their emotional memories. Loving touches make little lovers—or the lack of touch destroys them. Loving touches lay the foundation of health and emotional stability. They build trust and security. In a million subtle messages, decipherable only through the skin, loving touches promise the good life.

For information on how to find infant massage trainers, see "Resources," page 189.

AND THEN THEY GROW
(Touch and Young People)

> But I never seed nothing that could or can
> Jest get all the good from the heart of a man
> Like the hands of a little child.
>
> JOHN HAY

In comparison with early childhood, little research has been done on touch during the middle and adolescent years, but most knowledgeable parents and teachers swear by its continuing importance. The little girl who comes joyfully to sit on her teacher's lap as she shares her "show and tell" is unabashedly receiving strength from the encounter. The eight-year-old star of the Little League team, all rough, tumble, and macho, is among the blessed if he still feels free to climb in the lap of mother or father for a dose of security, a touch of extra strength, the spirit he needs to meet the challenge of the metamorphosis from child to adolescent.

Dr. Aaron Katcher of the University of Pennsylvania has talked about "idle play touch." It is that half-in-reverie, part mindless, part inattentive touch of animal or friend or object that seems to relax as nothing else can. A woman staring into space, one hand on a dog's neck, gathering up fur, rolling it around in her fingers, dropping it, picking up another strand; a little boy, Dr. Katcher's little boy to be precise, sitting on the front steps of his house, his leg and trunk pressed against his best friend, also sitting there—a spiral of his best friend's hair being twisted around and around his own index finger, and both children staring at cars in the street "without really seeing them, content in their mutual worlds, safe with each other yet free to let their thoughts roam a private internal world." This kind of idle touching, says Katcher, produces the same physiological effects as medication and can relax people enough to effect a significant reduction in blood pressure.

If touch, even idle touch, provides such peace for children, why not often use purposeful, intimate touch to comfort in very specific ways? Cuddling and stroking a sobbing child is the ultimate cure-all. Rocking small bodies against one's own body is pure nourishment for both bodies. When families are in the habit of touching out, both idle touch and directed touch come naturally.

If children don't feel satiated with the quality or amount of touching they're getting, they often revert to forms of self-touch. The little girl who twirls her hair endlessly or sucks her thumb obsessively is comforting herself because there's no one else around to touch her. Self-touch, in moderation, however, is absolutely normal—and that includes masturbation. One little eight-year-old was so comforting himself one day, when his mother caught him.

"You'll go blind if you do that filthy thing!" she shrieked.

The little boy stopped, thought about how good self-touch felt, and asked, "Well, can I do it just till I have to wear glasses?"

Touch cut-off time arrives very soon and it's time to "wear glasses" and stop self-touching and touching others, or so our society says. While it's fine to keep touching puppies or babies, many parents find it unseemly to keep touching their children past a certain age. "Sex rears its ugly head," as one parent grimly put it, and the most innocent of touch, for some, begins to take on distinctly sexual overtones. Touch takes a holiday—much to the silent dismay of recipients.

Puberty makes a difference to parents as well as children. The daddy who roughhoused happily with his son, kissing him buoyantly and unrestrainedly, suddenly starts to become nervous doing the same thing with a kid whose size is fast approaching his own and whose muscles seem distinctly more vigorous than his. As for his daughter who wants to creep onto his lap at fourteen—forget it. She has breasts! It just wouldn't be right.

At different points for everyone, no-touch rules take over. First they begin with warning our toddlers not to let any strangers touch them. Then, sadly enough, we warn ourselves that we also must cease the pleasurable activity of touching because it might be misread by others. And therein lies the irony: the desire for touch is never out-grown. There's nothing babyish about wanting that warm, sustaining comfort, but Western society in particular puts blinders on our needs.

School days may mean the beginning of many splendid things but they generally signal the slacking off of touch for a youngster, reports researcher Mark L. Knapp. The day a boy steps into kindergarten and gets his very own cubby, he starts to give up a whole lot of cuddling, and the pattern increases through the sixth grade, although he's still being touched more than he will be as an adult. When he arrives in junior high, he receives about half the touches he did in the primary grades and the touching itself is different—more the shoulder-to-shoulder or elbow-to-elbow variety rather than the hand contact he used to enjoy. In adolescence, if he's lucky, his touching picks up again, but only with members of the opposite sex. He treats his family and his same-sex pals as though they had some terrible disease when it comes to physical touch. There's no question that children and teenagers in the U.S. in particular turn away from their families for the good touch at a certain period in their lives.

If they've been lucky in the past and have shared much good and loving contact with those families, they will be able to transfer the loving touches to others. If contact has been minimal, they may have a hard time developing the physical closeness necessary to emotional intimacy with another.

Dr. James A. Prescott's research at the National Institute of Child Health and Human Development goes even further. The lack of early TLC—what he calls "somatosensory deprivation"—makes for later tragedy. "I believe," he says, "that the deprivation of body touch, con-tact and movement are the basic causes of . . . depressive and autistic behavior, hyperactivity, sexual aberration, drug abuse, violence and aggression." Warm, affectionate touching relationships in childhood or adolescence, says the developmental neuropsychologist, deter violence in adults.

The sulky, non-communicative teenager is not a myth—she exists. What happened to my adorable, cuddly Kim? thinks one mother. Well, many experts are not at all surprised that when the touching stopped Kim's communication with her parents became almost nonexistent.

Researcher Dr. Elizabeth R. McAnarney also wonders if the fact that parents and children touch so much less during adolescence somehow contributes to their difficulties in communicating with each other. "Adolescents need touch to facilitate communication," she says. "When children are no longer held and comforted by their parents, they may turn to their peers, instead," says Dr. McAnarney, who is director of adolescent medicine at the University of Rochester Medical School. "There is almost no data on this, but I wonder if the increase in very young teenage pregnancies comes from the need to be held. They may be using sex for a nonsexual purpose," comments the doctor.

Touch is a motivator. I have been to hundreds of parent-teacher conferences, and if I've heard it once, I've heard it a thousand times: "He's so smart, my son; he just lacks motivation." The parent is trying to tell the teacher that an energizing force is lacking in Sammy's life.

Well, I have a suggestion. More touching at home and at school. Touching is an energizer.

Children and adolescents come to respect the word more than the touch. Learning becomes intellectual rather than visceral. We say "hello" instead of hugging; we throw a sardonic phrase instead of a punch. In high school cafeterias all over America, political discussions take the place of pushing and pulling matches. And that's as it should be. If we are to have a safe and civilized society, our children must learn to substitute words for some actions—especially those prompted by anger. And then they must learn to modify their words so that these don't lead to aggression from someone else but instead foster communication and understanding. But our not-so-intellectual bodies still crave touch—touch of empathy and understanding. A teacher who tries to motivate only through words is missing out. As a teacher, I always indulged in the affectionate, not patronizing, back pat.

Strokes and gentle squeezes don't hurt, either. Studies have shown a distinct correlation between self-disclosure and touch. Nothing ever opened up a shy or troubled student as quickly as a loving stroke on his back or arm. It's almost as if a teacher turns a key to unlock hidden problems when he allows himself to reach out to a student. Conditions, of course, must be appropriate. One can't hug a kid in front of the class or in front of his girlfriend. One should also be watchful for the child who's uncomfortable with touch. You'll know him in a minute, if your antennae are out: that flinch or pull-back gives him away instantly.

If a teacher or a parent feels free to touch someone who enjoys touch contact, such rich communication is possible! People tell you secret things about themselves when you touch with a hand as well as a thought. But touch must be welcomed, not imposed.

Alas—often the most common touch scenario in childhood is touch as direction giver (hold my hand while we cross this street) or punishment (don't do that!). In adolescence, touch is usually relegated to aggression in sports or to confusing sexual messages. What happened to the tender touches of communication? What became of touches for empathy or understanding? Gone, gone. Tickling often becomes the closest thing to tender. Adults, frightened of having their touches misunderstood, simply stop. Pity. As young people move through childhood, then puberty, they desperately need the contact. Tender touch is a reassurance that they're not awkward, pimpled, ugly ducklings. Touch tells them they're as beloved as their cuddly infant-selves were.

I remember Eric. He was seventeen, in my English honors class, and he harbored a secret that was terrible for him. Theme after theme on the subject of homosexuality came pouring from his pen, "fictional" studies of one young boy's love for another. There was no question he was reaching out. He was shy and seldom approached me at school. One day I was amazed to see him pedaling his bicycle to my home.

Small talk ensued: the school play, the football team, college prospects, and then—the zinger. Eric confessed, shuddering, that he was homosexual. That he thought so did not come as a great shock to me, given his writings.

"But why, why are you so sure?" I asked. I took his hand in mine.

"Because," Eric blurted, "because I get turned on whenever Frank is near. Whenever he brushes past me, whenever he touches me, I die."

Frank was the prime high school jock. Cute, blond, tall—*I* was affected whenever he was near. So was the principal, distinctly heterosexual. He had charisma, Frank did, and was a charmer. But Eric, a naïf, told himself that being touched by a male peer, and liking it, was irrefutable evidence of being gay. At seventeen, he was certain he had no hope in heterosexuality. Brooke Shields was not for the likes of him.

I'm loath to be accused of oversimplification, but I *know* that taking Eric's hand released his mouth. His worst fears came bubbling to the surface. The fact was, he was terrified of touch because he wanted it so much. And, because he responded so intuitively to the touch of a charming peer, he was certain something had to be wrong with him. Whether or not Eric eventually found more solace in male than female arms, I don't know. What I do know is that a confusion about touch made him dig himself a neat little box, all prematurely labeled "different."

However, one point that should be reiterated: there are some people, many people, particularly adolescents, who have built "no touch" walls around themselves. It's a mistake to attempt to force touch on anyone who really doesn't seem to welcome it after your first overture. Resistance, discomfort, or active aversion to touch must be respected. There is nothing more intrusive than a well-meaning toucher touching a person who silently screams with dismay at the contact. How do you hear a silent scream? The look of disgust in someone's eyes might be a first clue.

If ever a generalization could be made, though, it's this: *try* touching. If it's unwelcome, you'll know it soon enough. Chances are it will be a source of support in times of joy and a source of inestimable strength in times of youngsters' stress. No one knew this better than the Reverend Dan Messier of Concord, New Hampshire. Rocked by the tragic death of schoolteacher astronaut Christa McAuliffe, residents gathered on the night of the tragedy to pray. Adults as well as strangers wept openly, hugged each other, and looked to God for answers. But answers were not forthcoming for a child named Tanya Lee, who lived in the Rumford Home for Girls, a local orphanage. She clutched a motherly woman she barely knew—and even that touching wasn't enough. The Reverend Messier spotted the youngster sobbing uncontrollably. He stopped the service, left the pulpit, and went down among the parishioners to take the orphan aside, hug and hold her.

"I love you and I care for you," the priest told Tanya. "It will go away, just hold on to me—tight." She held on. Slowly, the tears stopped.

So much for the power to touch young people, and for the infinite wisdom of some adults to understand what touch can accomplish.

DO YOU LOVE ME
FOR MY BODY OR MY MIND?
(Touch and Sex)

Do you love me for my mind or is it just my body?
Just my body? If you're smart, you'll hope it's definitely for the

body as well as the mind. Although intellect is nice, body is your basic terrific. Intelligent people should pray that what Emerson dubbed "the electric touch" enters a relationship. It's no accident, by the way, that the metaphor reads "electric"; touch galvanizes bodies, makes poetry of them. Touch arouses places you never knew were sleeping.

"Touch not, taste not, handle not," warns the New Testament. No disrespect intended but—ignore it. Surely any beneficent God would not forbid that which He must have given a lot of thought to—touch as trigger to great love-making. Sexual touching is one of His most inspired inventions. And all good sexual touch is sensual.

A sensual touch is one that gratifies the senses—all the senses. It is voluptuous. It is slow, searching, and attentive to reactions. It gives and receives. It is enjoyed as much by the giver as the receiver. Sensual is related to sexual—can lead to the other, but it's not necessary for it to do so. Sensual is glorious in itself. Wise lovers know that sex does not necessarily mean intercourse but rather sharing the exquisite delight of exploring one another's bodies. A sensitive toucher can evoke wonders. With sensual fingers, a partner's touch can instantly cool a head and eyes that are burning with fever. With sensual touch, lovers can instantly gentle a damaged psyche as they spoon-fit off to a calming sleep. And then, when the time is ripe, he or she—or he *and* she—know how to transform the sensual touch into the sexual. Fingers and hands, which have practiced the craftsmanship of sensuality, can touch your libido, make it surge with desire, make it burn brighter and steamier than you ever imagined possible. The sensual touch that nourished the infant, expanded the language, comes into glorious fruition during erotic play.

Masters and Johnson and the many sex therapists who use their methods allow no intercourse at all in the first few weeks of sex therapy. Couples practice touch—gentle, appreciative, sensual touch of one another's feet, arms, legs, face, hands . . . and bodies. They carefully learn the art of touch as a wondrous process in and of itself. Intercourse is *not* a goal, they teach. If it happens because both lovers are in the mood, fine. If not, sensual touching is golden anyway.

Intercourse "as a goal" seems to be a big hang-up in our culture. Most sex therapists are busy warning against goal-oriented sex as a sure way to decrease sensual enjoyment. In a fine manual, *Touching for Pleasure,* Susan Dean, a specialist in human sexuality, and Adele Kennedy, a professional sex surrogate partner, advise: "When you are goal-oriented you are no longer in the present. You interrupt your responses when you try to anticipate the next step. Stay focused in the moment, so that you receive all that is being offered. Touch is, in and of itself, the pleasure. Any by-product is coincidental to the experience, although it is to be enjoyed and appreciated. The motivation is to sensitize your body to its intrinsic capacity for pleasure."

The eighties guru of making love, Alexandra Penney, does not disagree. Great sex, she states baldly, is between two people not two bodies. It goes far beyond mechanics; there are all kinds of great sex besides intercourse.

Whatever your kind is, it can't happen without touch.

I take it back.

I remember Marlon Brando's *Last Tango In Paris*. (In Latin, incidentally, "tango" means "I touch.") Brando sat on one side of the room, his lover on the other side, and they made passionate love with their eyes. The actors were so aware of each other's body rhythms and timing requirements that the audience felt exhausted and damp with their love-making. And, in a Broadway show about two imprisoned homosexuals, the actors made love with only words and gestures, so poignantly that one would swear they were touching each other's flesh. As they reached a mutual onstage orgasm, without having actually touched once, the entire audience gasped with shared relief.

The reason that both the movie and the play succeeded is that the simulated touching was so sensual the audiences literally felt touch pervading their own beings.

We are sexual creatures. Normal infant baby boys masturbate happily. The elderly—those who are not turned off by society's stingy expectations of them—are also sexual creatures, though they might not practice intercourse quite as often as they once did. We are sexual because we are sensual and, once having experienced the pleasure of body electricity, we can anticipate it and even simulate it in our brain. Ever feel a rush of flutters in your groin while reading a sexy passage? You *remember*. Touch is so powerful it lingers in our bodies even when it's not actually there. That's why audiences flock to superbly written sexual scenes on-screen or in print.

The skin, the medium for touch, is such a wondrous conductor of sensual signals that it changes either visibly or in ways that can be felt when it is touched, when it anticipates being touched, or even when it remembers being touched. How many of us have felt an actual warming of the skin, a literal heating-up at the very thought of a single touch from a special person?

Many primates have what researchers have called "sexual skin"— the mandrill, for example, with its blatant coloring consisting of red and blue rump patches. Some monkeys have their sexual skin around the

face, some around the anus, some around the vulva. The skin begins to
swell with the beginning of each female cycle and reaches its glorious
peak of color and fullness just when ovulation occurs; the skin may be
so swollen at that time as to incorporate one-sixth of the animal's total
body weight. After ovulation, the swelling recedes as fluid is passed
through the urine and the color recedes. Although the human female's
skin is not so blatantly telltale when she is feeling most erotic, it also
warms and color-heightens.

Desmond Morris, in his classic study of man, *The Naked Ape,* really
gets down to the nitty-gritty as he discusses touch as a sexual contact:

> After the initial stages of visual and vocal display, simple body con-
> tacts are made. Hand-to-hand and arm-to-arm contacts are followed by
> mouth-to-face and mouth-to-mouth ones. Mutual embracing occurs, both
> statically and during locomotion. . . . Body-to-body contacts are increased
> in both force and duration. Low-intensity side-by-side postures repeat-
> edly give way to high-intensity face-to-face contacts . . . [V]arying pres-
> sures [are given] from all parts of the body, but in particular from the
> fingers, hands, lips, and tongue. Clothing is partially or totally removed
> and skin-to-skin tactile stimulation is increased over as wide an area as
> possible. . . . [T]here is a good deal of skin manipulation. The hands and
> fingers explore the whole body surface. . . . [T]here is also a considerable
> amount of twining and inter-twining of the arms and legs.

All that twining and intertwining that the naked ape goes through,
when described scientifically, can sound lethally dull. But, there's more
to sexual touch than scientific descriptions allow.

There's the brain, for example. Woody Allen says that the brain is
his *second*-favorite organ, but the brain is really quite primary when it
comes to sensual or sexual contact. It interprets touch, and tells you
whether you should let a caress be enjoyable or despicable. Your father
can run his hand along your arm, an ax murderer can run his hand along
your arm, the love of your life can run his hand along your arm, and the
touch that differs not at all in intensity or quality will feel entirely
different in each situation. Your brain filters the information and tells
you how to react. If your brain tells you that the woman in front of you
is your dream of an ideal sex object, you will enjoy touching her. On the
other hand, if your brain tells you that the woman in front of you is too
old, too young, dumb or smart, dirty or diseased, touching her will not
be enjoyable—despite the fact that her skin may be buttery soft and
smell like fragrant frangipani.

Not that you have to love someone to respond to sexual touch. As

Mae West put it, "Honey, sex with love is the greatest thing in life, but sex without love—that's not so bad either." So you don't *have to* love someone to react well to a sexual touch. Technically, you don't even have to like anyone to be excited by contact, as is shown by the myriads of anonymous and still reasonably satisfactory sexual contacts made by men and women. So erotic and so sensitive are the erogenous zones that if you look forward to being aroused, if you concentrate on responding to sexual touch, if you *think* you'll be turned on, odds are you will. After all, people are the sexiest primates. Females in other species are limited in sexual receptivity to times of ovulation but not the human female, who just loves making love any old time—*if* she's in the mood. And that goes for all ages after puberty—including old age.

Still, as the sex therapists point out, it helps a whole lot when caring is present. Tenderness, respect, and practice have a way of intensifying physical pleasure. Knowledge of an individual's needs and desires—and respect for them—can be paramount. I don't mean to overlook the kick of errant or forbidden sex—especially as adolescent adventures (at any age!)—but most of our experience, as well as our literature, music, and drama, teaches us that sincere, knowledgeable sexual touch can give the heightened sexual pleasure that a one-night stand can never attain.

If touch doesn't come from a place of love, it simply doesn't work so well. But, when it does, it has almost miraculous results.

Listen to what happened to sex therapist Adele Kennedy, who had a long-time touch-sensitive connection with her lover:

"[He] had arrived at my house one night to take me out to dinner and a film. We sat down in the living room to discuss which restaurant to go to. As we were talking, he was running his fingers gently over my knuckles, and I was looking through a magazine for dining suggestions. Before I knew it, I had an orgasm. . . .

"We were both surprised, as there was nothing outwardly conducive to it. Although our conversation or nuances had in no way been sexual, we were connected; and since I had allowed that connection to flow freely, my body responded long before sexual thoughts ever reached my consciousness. It was at that point that I knew that the possibilities of stimulation and response were infinite."

And personal love seems to breed love and compassion in a wider arena.

Murray Strauss, a sociologist at the University of New Hampshire, who has spent his career studying family violence, and his colleague, Roger W. Libby, conducted some research on college students and found that those who engaged in sexual activity as a warm, affectionate bonding committed fewer violent assaults. Those students who engaged in sexual practices as a power struggle—as a domination or an unaffectionate kind of scoring—were more apt to be involved in violent assaults. Not too surprisingly, male students who thought of sex as a warm, affectionate bonding were also in favor of gun control.

I'd like to note that, while the men were at the high end of the dominant-competitive scale, women were at the high end of the warm-affectionate scale. That's because, for one reason, girls are far more likely to be touched from birth on than boys. And, while this may make them warm and affectionate, it also makes them tend to take what they get, rather than to be assertive in the getting. Women learn to let themselves accept sexual touching rather than initiate it. They are believed to be more passive sexually, and thus they train themselves to be so. They worry that men will find them pushy, unfeminine, "grabby" if they are sexually aggressive with touching. Boys, on the other hand, learn to be the touching aggressors. When they are men, they often have little patience for lying back and receiving touches in foreplay. Lie back and just enjoy? Is that really manly?

Although this attitude seems old-fashioned, it still persists in spite of many studies and books to the contrary. Although our generation is supposed to be one of sexual sophistication, insidious double standards continue.

Touch is then relegated to a sexual provocation instead of being a mutual and enduring pleasure in its own right. If one doesn't love to experiment with touch as a conveyance of affection and friendliness, one may never reach the zenith of sexual achievement. Sensual touches must precede sexual intents for both men and women.

A young man may place his hand on his girl's breasts two seconds after they tumble into bed. Or perhaps he heads straight for her genital area before she has time to catch her breath. He expects that once he touches these sensitive zones her sexual motor will automatically shift into high gear.

But of course it doesn't. She feels annoyed, outraged, or, at best, nothing. Her motor remains in idle position, or perhaps it stalls, chokes, and turns itself off.

He feels confused or nervous. Is there something wrong with him if his touch doesn't spark a conflagration? He may feel irritated; what a cold potato his partner is! It's not his fault, surely.

And therein, note Masters and Johnson, lies a basic dead-end approach for many relationships. He thinks she's frigid because either she didn't let him touch her in the crucial place or, when he did, she wasn't thrilled. She decides he's a boor and a lousy lover. They learn to settle for mediocre. They consider touch merely as a means to an end. They buy sex manuals that teach them how to manipulate body parts—"the science of stimulation" of disembodied parts. How much better to have a philosophy that celebrates caring, that celebrates sensual touch as an extension of everyday life. Sensual touching should not be chained to the bedroom. It should not be the exclusive property of the male half of the species.

The bottom line is this: we—both men and women—miss sensual touch. We confuse it with sexual intercourse and so we deny its casual use. And then, ironically enough, we separate it from sexual touch. One cannot be the other. Wrong. One is the other.

And so, when we climb into our sexual beds with our lovers, we often don't understand that cuddles belong there as well as "G" spots. Even some otherwise sophisticated people consider the urge to be nurtured and stroked a childish urge. Curling up, in a "spoon" fit, holding one another tight and warm is an immature fantasy in their lexicon. It's infantile, they say, to want a snuggle in bed instead of the real thing—hard, passionate sex.

Masters and Johnson pointed out that too many people interpret every sensually loving touch as an invitation to copulate. Bodies hunger for holding as something quite apart from genital penetration. Dr. Marc H. Hollender, a psychiatrist with the Department of Psychiatry at Vanderbilt University School of Medicine in Tennessee, has said that the need to be held is so compelling for some women they often barter intercourse for being held.

"I don't warm up to words very easily," said one, "so being held has a lot of significance for me."

"The need for just touch is like a kind of ache," said another.

"I'd rather have my husband hold me than have a Cadillac convertible," said yet another.

"The reason I like to wear big, fuzzy sweaters and be bundled up and held warm in blankets," said one, "is because it makes me feel like I'm being held."

If there are men who don't know that sensual touch can be a sexual dream, all by itself, they should be gently told, says Dr. Hollender. Some women tell by words. Others tell by actions.

Touching fuses and binds. It can begin the heady sexual dance, but sometimes it *is* the heady sexual dance, all by itself.

"These days," said one woman, "you just can't risk putting a hand

on someone or leaning back into his arms as he helps you with your coat.
All you want is a loving touch and it's interpreted as something else
entirely."

The trouble is that many of us wear our hearts not deep in our
chests but right there, on our skin. And hearts read touches too impa-
tiently and often wrongly. Sometimes we're not able to see sexual touch
as a continuum. We're too used to turning ourselves off during affectionate moments, says sex therapist Dagmar O'Connor, the director of the
Sexual Therapy Program at St. Luke's Roosevelt Hospital Center, in
New York. As teenagers, many of us felt we had to turn ourselves off
when we hugged and kissed our parents, and we do the same thing now,
with our own children. But we can enjoy the nurturing of the simple
hug and kiss with our parents and friends if we remember that these
are exactly the same physical acts that can turn us on during actual
intercourse. They're good all by themselves.

"When you are making love," says O'Connor, "you should experience the pleasure of touching rather than watching for a reaction
. . . many men are stuck in the ego trip of touching a woman and gaining
pleasure from watching her being turned on instead of forgetting about
the woman and getting into the experience themselves."

Many men and women are stuck, as well, in the myth of the "erogenous zone." There are only a few erogenous zones, reads the myth.

On the contrary, people who spend a lot of time thinking about
erogenous zones know that the erogenous zones are legion. There are
the genitals, naturally, and the mouth, and the inside of the ear, and oh,
have you ever had your palm and fingers caressed? Or your feet? Savvy
lovers also understand that touch incites myriads of body zones they
would only know about if they spent time, lots of time, in mutual
experimentation. A careful lover, stroking a cheekbone, a chin, a brow,
can lead herself and her partner into a sexual frenzy. Feet, especially
in between toes, are delightful for sexual touch. Every patch of skin is
a possibility when affectionate exploration is built into the touching,
when sensual feeds the sexual and time is spent savoring both.

Taking a look at the traditional erogenous zones can be quite interesting. The kiss, for instance, is uniquely human. No one else really
kisses even though guppies and other species look like they're doing
it. Americans kiss no matter where they live, but up until twenty-
five years ago there was no word for kiss in Japanese and native
Chinese didn't kiss, ever. Kissing was not originally seen among the
Australian aborigines, Maoris of New Zealand, Papuans, Tahitians,
South Sea islanders, and Eskimos, although low air fares have

imported the practice to all of these places.

Oral and manual stimulation of genitalia are a whole lot more universal than kissing. Only humans use manual stimulation of the breast as erotic arousal, but manual stimulation of genitalia is both practiced and enjoyed, in other species, internationally.

In many world societies painful stimulation is a touch turn-on. What is known as sado-masochism is familiar in America but considered fringe sexuality. On the other hand, painful touch in some other societies is sexually "normal." The Apinaye women of Brazil, for example, bite pieces of their partners' eyebrows off and spit them out loudly. The Siriono of South America may poke at partners' eyes, while they scratch and pinch at each other's skin. Trukese women of Oceania just love ear poking, which seems to excite their lovers inordinately, and Trobriand Islanders bite each other on face and lips until blood is drawn.

Human beings experiment with sexual touch in infinitely varied ways. The more they practice touch, the more they know what feels good for them.

Knowing yourself and your needs—aside from knowing the specific erogenous zones—often tells you *when* the sexual touch is most appreciated—most needed. In our fast-track society, we often receive as well as give double messages of such complexity that we ourselves are confused. What feels like sadness can often be a need for the sexual touch. What feels like fatigue or hunger can be a disguised wish for love-making. Although this is a complicated point, not-so-complicated television messages are being developed in the newest attempt to move Americans to get in touch with their emotions and bodies.

The camera, in one such new spot, focuses in on a pretty thirtyish woman as she looks over the contents of her refrigerator. Rejecting the ice cream (too sweet), she samples some leftover Chinese food. Not satisfying—she doesn't know why. She closes the refrigerator door, then reopens it. Her dissatisfaction is familiar.

Finally she says, "I have the entire contents of this refrigerator memorized. I hate myself for this. I hate my diet. I hate my body. I hate my boss, and I'm freezing to death! But I sure would like a piece of fudge."

Then she backtracks. Gets thoughtful. "That's not true. What I'd really like is a hug. And I'm not going to find it in here."

The commercial, designed as a slice-of-life prelude to the pitch by American Cyanamid that shortly follows, unerringly manages to hit on the loving-touch deprivation that is so prevalent that an advertiser expects millions of women to recognize and identify with it. There are, thinks American Cyanamid, a whole lot of potential customers out there whose responsiveness to sensual touch is so blunted by the modern world they don't even know it when they need it.

Responsiveness to the knowledge that one *wants* sensual touch, as well as responsiveness to the touch itself, is fragile. We must retrain ourselves into recognizing and accepting our basic desires.

Still, our present era has taken giant steps forward since our parents' and grandparents' generations. No such ad would have been possible then; nor would anyone have been attuned enough to touch to dream it up. And, surely, no book like *Touching for Pleasure—A Guide to Sensual Enhancement* would have been allowed on the shelves of proper bookstores, especially with its highly explicit drawings of interacting human bodies. And that's not even to mention Dr. Ruth Westheimer, bubbling along as a petite, white-haired grandmother on prime-time TV, asking questions like "Was she moist when you penetrated?" Can you imagine that happening twenty years ago—or even ten?

In "the good old days," many a husband and wife never saw each other naked; all those nightgowns and pajamas must have been a dreadful impediment to the delicious sensations of the skin. Also, there were no-nos about touching certain parts of the body—and for some there still are. Sixty years ago, one certainly never spoke about sex with one's husband. Heaven forbid! And, although their stomachs were nine-months visible, women never mentioned their pregnancies—often even to their very best women friends.

Many say that today's society is obsessed by sex, pornography, eternal searches for youth, and advertising that is predominantly sexual innuendo. But the attitudes reflected were present during the Victorian Age; they are simply more visible today. So too are the realistic appreciation and understanding of the sexual needs of human beings. It is my contention that as sexual and sensual touching grow more respectable and are talked about in the open as something that decent people actually do at *all* ages, then pornography, sexual kinkiness and abuses will decline. Dr. Ruth may be doing us all a great service in her grandmotherly acceptance of the sweet touches of love—and her refusal to give advice about love when it is used destructively.

A sweet touch is a touch that evokes anything from giggles to cuddles to goose bumps in a partner. A sweet touch always provides a personal and sensual sense of well-being. It never makes one feel used. Sweet touch, true sensual and sexual touching, is never destructive.

Actually, it's glorious. And perfecting it can be a life's work.

DON'T STOP NOW
(Touch and Growing Older)

Touch us gently, Time!

BRYAN WALLER PROCTER

My mother lay in her bed at the Rusk Rehabilitation Center, the victim of a stroke at seventy-six years of age. She was terrified that she was going to die. Her blood pressure was alarmingly high, but the doctors did not feel it was wise to medicate her any further.

All day long we held her, my husband and I. Her grandchildren stroked her. Her condition was stable. Night fell and we had to go home. She tried to be brave and she smiled and sent us off to our lives.

In the night, her fear became worse. She felt dizzy and weak and her heart beat in her breast as if it were a wild thing. She was certain she'd never see daylight.

From somewhere in the night came a young nurse.

"Please touch me," said my mother, instinctively knowing what she had to have. "Please, can you hold my hand?"

The nurse sat there for an hour holding the limp hand.

My mother's heartbeat slowed, notes the hospital record. Her blood pressure dropped to near normal. If she didn't feel well yet, she was calmed. She mentions that nurse probably twice a day, and she has done so for the last six years since the incident. To hear her tell it, the young nurse saved her life with touch.

The older you get, the more you need to be touched, but ironically enough, the less people want to touch you. Baby skin is smooth and fragrant—squeezable. It's soft to the touch. Aged skin is also surprisingly soft but brittle and thin as well, cracked- and parched-looking.

"He looks so old," said a little girl at the hospital to her mother when she was urged to kiss her grandpa. "I'm afraid he'll tear if I touch him!"

From the mouths of babes often are heard the truths that silently shout in the hearts of grownups. In the *American Journal of Nursing* is a study from Sara Jane Bradford Tobiason, R.N., M.A., who is an assistant professor teaching gerontology and geriatric nursing at the Arizona State University at Tempe. She found that baccalaureate nursing students voiced the same concerns as they worked with the hospital aged. Tobiason suggests that anxiety about touching the elderly may stem from fearful attitudes about what the aging process means to the touchers themselves.

The researcher tested groups of students and then queried them after each had spent time touching both newborn babies and the hospital aged. After touching the babies, the students felt an increase in "protective" feelings, but after they touched the aged, they felt a decrease in "loving" feelings. One nurse, echoing the little girl who was afraid to touch her grandpa, said she was certain her patient would "break" if she touched him. The newborns, said the nurses, felt "cuddly." The elderly felt "dry."

Despite what they felt like to the hands of the nurses, the aged were as reliant on touch for nurturing as any one of the plump, gurgling babes. The need for touch does not increase with age, but neither does it decrease, and amazingly enough, when every other sense is failing and driving grandma up a wall with its inconsistency, touch remains relatively consistent in the accurate messages it gives. In fact, up to age fifty-five, touch sensitivity remains exactly the same as it is for the young adult, and after that the touch threshold rises comparatively slowly. The hands receive the greatest attrition, with the fingers and palms becoming somewhat callused, with less ability to give and receive communication by touch, but the rest of the body remains extraordinarily receptive to caress.

What does change with age is the *opportunity* to be touched and to touch. Many of our senior citizens have lost spouses and siblings, and their children and grandchildren—those who would be most likely to give tenderness by touch—have long since fled the nest, the state, and, in many cases, the country. Even if the elderly are still functioning in the real world, as opposed to being confined to homes or nursing facilities, few are left to touch them. Even politicians who eagerly kiss and hold babies tend to eschew affection when they meet the elderly, who, unlike babies, hold a valid vote. The vigorous handshake the politician has developed to show his virility is shrunken to a limp touch as he also parades his fearfulness that the elderly might break.

As the other senses fail, the elderly need touch desperately to keep them in the real world. Information is filtered in and out, cues and clues to myriad new ways of doing things are perceived more and more through touch as it now enhances failing sight or hearing.

One of the major problems of aging, says Tobiason, is a sense of alienation and "disengagement" that older people experience. Touch, she maintains, is essential in giving them a sense of involvement, in helping the elderly to interact with the rest of the world, which seems to be speeding by on twenty-three-year-old feet. Touch stimulates. It wakes up. Simple as that.

Joan Mowat Erikson is a California-based artist-teacher-researcher who has done extensive work with the aging. Herself over eighty, she belies the image of "old woman" that society would turn her into if it had its druthers. Instead, Erikson is long and lithe (despite a hip replacement), immersed in work and the future, and worrying about the lack of touch in the lives of her peers.

"The skin, which from head to foot relates us sensitively to the world in which we live, our matrix, is indeed our most consistently active and informing organ of sense. In a dark vacuum," says Erikson, "where only minimal sight, hearing, taste, smell and muscle activity would be possible, the skin can still report something of the nature of the surroundings."

Reports back, skin does, and Erikson maintains that touch is teachable and those who do not receive the daily touching they once enjoyed can replace it with "formal" touchings like daily massage, which not only replaces the missing touch stimulation but relieves tension and "loosens tight places." Dance is another touch form that the elderly can enjoy. "It gives an excuse for touching, makes body—not just finger—contact."

There are boundaries that must be respected, says Erikson, when approaching the elderly. "Someone who has the defensiveness of a porcupine would bristle if you touched her when she was twenty and also if you touch her at eighty. Defenses that are life-long are even more firmly entrenched as one ages and one must never intrude on the autonomy of an older person as if she were an infant."

Still, there are few elderly who don't respond to warm touch. The trick word in that sentence is "warm." Hospitals and nursing homes are often sterile and bleak. If medical attendants touch inhabitants, the touch often comes in the form of a condescending pat, a forceful poke, an examining probe. We must somehow change that kind of touching, says Erikson, and teach loving touches of our patients who, in the main, are tender and responsive beyond belief to such contact.

Basic routines to care for the elderly must be "rethunk," maintains Joan Erikson—and wheelchairs head the list. Wheelchairs, she says, are an abomination and the nemesis of touch. For one thing, they provide barriers on either side of an eighty-year-old who is wheelchair-bound: when two wheelchairs are side by side, the inhabitants are literally barred from touching each other. Wheelchairs make jails. People have to stoop to speak with you, as if you were a child. And the wheelchair-bound, because they must strain to hear the words of a standing person, and strain to touch the face of an upright attendant, and strain to be part of the world of store counters, elevator buttons, and everything else the wheelchair makes unreachable, are in jail. For starters, says Erikson, reinvent the wheelchair, and then think about how to surmount the other fences, all the barriers, which include walkers, geri-chairs, and raised sidebars on beds, which some elderly need as assistive devices.

Older people must be encouraged to think about their relationship with materials as well as other people. One can relish the feel of the fork one's just picked up, the surface of the table, the velvet of a scarf; touching pleasures are boundless. As a sculptor sensuously molds his clay, the elderly can learn to be sensitive to the touch experiences that are available—even if they tend to eschew being touched by other people.

Maggie Kuhn is the founder and leader of the Grey Panthers, a group that is militant about the rights of the elderly. Touching prejudice is rampant, she says, and if ever ageism reared its ugly head, it's in this area. Older people do not break from a touch; on the contrary, she says, "We thrive when our touch hunger is satisfied . . . as does everyone else!" The trouble is that when touch hunger is so often unsatisfied the elderly can easily forget how good it was when they were touched. In their great disappointment, they tend to stop asking—to become uncommunicative about their needs. It's hard to ask to be touched. When a casual brush on the cheek is given, or a handshake, or a condescending back pat, the result is often so "teasing" it's worse than nothing. Only a true, caring caress will do, says Kuhn; it's better to withhold the perfunctory, non-caring pecks than inflict them patronizingly on a touch-needy person.

In very old age, says anthropologist Ashley Montagu, the male's sexual capacity is diminished. However, his "tactile hunger is more powerful than ever because it is the only sensuous experience that remains open to him." Dependent on the support of his friends and family, he needs embraces—and the chance to give them back.

Women, once whistled at and hungered for as sexual creatures, are not now seen as women but almost as ineffectual objects. Where once they were cherished and held with pleasure, they are now perhaps still cherished, but held—if at all—with far less pleasure.

In America, unlike many other societies, the elderly are viewed as people "changed and lessened by aging," says hospice nurse Cathleen Fanslow, rather than as people still in the process of becoming. If the elderly are to keep "becoming," staying in touch with touch is more vital than ever. It acts, touch does, as a grounding—a safe and easy place to be—and the forgotten and the untouched elderly are dying before their deaths.

Our aging skin is exposed—it's public. Face lifts, moisture creams, and surgery aside, nothing can really make it fit as it used to. Grown saggy and baggy, leathery and splotched, it often precludes the holding we never for a moment stop needing. Estée Lauder, the tycoon of cosmetics, once said, "It's not aging that women mind so much—it's the look of aging." It's true. We become resigned to a circumscribed number of days, but few ever come to be comfortable with the way the years have changed their skin.

If the look of our skin changes, its job doesn't. It's still there to protect. It's still there to convey messages to the heart and soul.

The most urgent message is this: if you know someone who's touched and been touched with pleasure all her life, don't stop now. When cultural codes put up invisible signs that say, "This person is no longer cuddly—she's beyond touch," it is up to caring people to tear down the "don't touch" signs. One teenager, visiting her grandmother, put it wonderfully: "It's as if every time I see Grandma, a neon light begins to flash on her face. It says over and over, 'Please touch me, please touch me, please touch me.' So I do."

THE LAST TOUCH
(Touch and Dying)

She was old. She was dying.

She lay in her bed in a hospice and a nurse cradled her nearly bald head in her young arms. She sighed with a deep relief.

"If you only knew what just holding a finger can do when you know you're never going to be able to eat peanut butter and jelly sandwiches again," the old woman murmured to no one in particular.

When illness is profound, and there is nothing, no medicines left at all, then touch is still left and it is the most comforting medicine after all. It is a real medicine. It heals terror. It smooths away depression.

Anyone who has ever seen Elisabeth Kübler-Ross work, or who has read her books on death and dying, has seen the enchantment of the last touches. In an interview, Kübler-Ross once told me of the little boy who had to die and whose parents wouldn't let him go. He lay in his second-floor bedroom, a sickroom, and everyone visited, hands off most of the time, and everyone felt horror. His siblings didn't understand why they were isolated from him when nothing he had was catching. His parents raged against the dying of the light, and most of the time they stood at the door while he dozed, and cried.

Kübler-Ross came to visit, called by a mutual friend.

"Take him out of that sickroom," she counseled. "Take his bed to the living room—where you all *live*. Every one of you—eat on his bed, play on his bed, watch TV on his bed, touch him every moment you can. Let him live through your constant contact for as long as he lives."

They did it. The child responded, not with health but with raised spirits. He was part of life once more—if he was part of touch, if he was *in touch* with the life of the family. He didn't seem quite so weak, anymore, quite so fragile. He wasn't untouchable. Family life focused on the sick boy's bed in the living room. He was in the middle of life, in the middle of his brothers' game of catch, in the middle of his mother's telephone calls. No one was saying "goodbye"; everyone was living his own life around and with the sick boy.

Two months passed before he died, and after the family had worked through the terrible mourning, the mother wrote to Elisabeth Kübler-Ross and thanked her for giving back the touch of her child before it was too late.

The last touches are sometimes the most important touches, says Stephanie Matthews Simonton, who has made a life study of families who face grave illness. They are the touches that count the most to the sick person and are remembered the most tenderly.

"A person who is ill," says Simonton, "has an increased need to be held and loved." Like a child, he feels vulnerable and helpless. For the patient who hates the way he looks and feels, physical affection is one of the best ways to communicate acceptance and love.

Sometimes a touch allows the patient to cry—a restorative act. "Sometimes," says Simonton, "I've walked in to find the patient emotionally constricted, attempting to keep the family members around the bed from knowing how much he was hurting. I have often sat down by the bed and just held the patient's hand, perhaps stroking it gently —sometimes, not saying a word. On occasion, some of my 'toughest' patients have broken down and cried. The caring expressed by touch burst the dam and they were able to gain some relief from their pent-up feelings."

It is interesting to note that health-team personnel tend to touch patients in good and fair condition 70 percent more often than those more seriously ill. Being in a hospital, being *very* sick, puts a "Don't Touch" sign by your bed. You're penalized for dying, as it were.

Sad. The last touch counts desperately. The last touch makes it easier to leave life.

Cathleen A. Fanslow, R.N., M.A., is the Oncology Clinical Nurse Consultant for the VNS Care Program (Hospice) of the Visiting Nurse Service of New York. She teaches the nurses who train under her to use Therapeutic Touch (see page 66) for many illnesses, to help patients relax, to relieve their pain and reduce the swelling that frequently comes in cases of mortal illness.

And she does something else. Cathy has developed her own special touch procedure, which provides a comforting bridge for the dying to cross, a method that eases the way out of life. She teaches this to family members and to her home health aides so that they can help a person die peacefully. Listen to her describe it:

"The body, as we know it, is a dense, physical field and just outside it—from 3/4 of an inch to an inch and a half—is the bio-energetic field. The energetic field is intact when someone is basically healthy—except for some disturbances when a person is temporarily sick or injured. But, when a person is really ill and begins to die, this field becomes more fragile and friable. It actually starts to open up—to separate from the body. When a person is dying, she often feels abandoned by others and the family or the aide feels helpless. The touch protocol helps enormously in keeping a bond between the living and the dying and it gives permission to the dying to let go without fear.

"First you unruffle the bio-energetic field by passing your hands slowly over the whole body about an inch or so above it. This smooths the sensitive energy field and calms the patient. Then you move to the patient's left side, the heart side. (There is a direct relationship between the hand chakra—the center in the palm of the hand—and the heart chakra. Both are energy inflow and outflow centers.) Very gently, you place your left hand—palm up—near the patient's left hand. Then, with your right hand, you take the patient's hand and place it in your resting left hand so that you are holding thumbs. Next, place your right hand over the patient's hand—gently but firmly cupping the patient's left hand with both of your own.

"Patients almost invariably say something like 'it feels as if you are holding my whole body.'

"Then, silently, you project thoughts of peace and love towards the patient," continues Cathy.

"After a while, you free your right hand and place it on the left shoulder of the patient and very gently push it down towards the heart; you leave your hand over the heart area, sending strong messages of peace and love. The patient seems to receive a sense of enormous warmth and security," says the nurse, "and knows that it's all right to let go and die."

Sally was a forty-year-old woman who, gasping for breath, was dying from lung cancer. Herself an orphan, she was terrified of depriving her fifteen-year-old son of a parent. It was a tremendous struggle for her; the idea of separation from her beloved husband and young son was awful. She woke up in the night with cold sweats and nightmares.

"We worked with Therapeutic Touch and visualization," says Cathy Fanslow. "She'd imagine herself in a safe glen and sometimes it worked but the fear was still holding her to earth."

"One day, I said to her, 'Please tell me how I can help you. I want to be really *present* for you.' The woman burst into tears; it was apparent that her need for emotional succor was profound.

"Her husband was having a terrible time letting her go, as well. I instructed him how to hold her hand and send messages of love and peace. As I watched, he suddenly said, 'ohhhhhh' and he seemed to change as he looked at his wife. So did she. She visibly relaxed. Soon after, her shortness of breath stopped. In a few hours, she died peacefully."

Afterward, Cathy asked the husband what had happened.

"I was holding her hand," he said, "willing peace and love into her tired body. Then, from nowhere, into my mind came the word RELEASE. I sent it to her."

Helen Keller said, "Paradise is attained by touch," and those who need the last touches, the last contact with the world, know well what she means.

If anyone you know is dying, gently touch her into that new place. Rub her, embrace her, or just hold her hand.

CHAPTER THREE

HEALING:
The Historical Healer

I firmly believe that if the whole materia medica *as now used could be sunk to the bottom of the sea, it would be all the better for mankind—and all the worse for the fishes.*

OLIVER WENDELL HOLMES

Physicians pour drugs of which they know little, to cure diseases of which they know less, into humans of which they know nothing.

VOLTAIRE

If you made a study of civilization's earliest healers, you'd find that their *materia medica* consisted primarily of hands. Oh, perhaps a bit of sacrificial blood or holy water was added, but hands, and thus touch, were the golden cure.

Nothing gives so clear a picture of ancient medicine as the Ebers Papyrus, discovered in 1872 in Thebes. It had lain in a tomb over a decomposing mummy for over 3,400 years, but was greeted with joy by its discoverer, Dr. Georg Ebers, who bought it from a not-so-crafty Egyptian. The ink on the extraordinarily rare document was as fresh as if it had been written today and the date on which it was written was fixed at 1553 B.C. In graphic words and drawings, the papyrus describes medical preparations and practices of the time. Egyptologists established that it belonged to Amenhotep, a pharaoh who ruled Egypt from 1557 to 1540 B.C., two hundred years before the famous Tutankhamen. The document describes, among other things, an Egyptian medical treatment that corresponds exactly with what healers today call the laying-on of hands.

In Babylonian medicine, surgeons who cut with knives were not looked upon with as much favor as the hand healers. In fact, cutting surgeons didn't have an easy time of it at all. The rule was that if a surgeon treated a citizen with a "metal knife for a severe wound" and cured him he would receive ten shekels of silver, but if the surgeon treated someone with that metal knife and the patient died the surgeon's hands would be cut off. Not much of an inducement to desert touch healing for surgery.

The ancient Greek god of healing was Asclepius, the son of Apollo. With his "god hand" and a simple touch, Asclepius went around not only curing the ill but raising the dead. He married the goddess Hepione and had five children, all chips off the old healing block, as their names declare. One of his daughters was Panacea, the goddess of the relief of pain, and another was Hygeia, goddess of health—you get the idea. Asclepius's touch was so potent that eventually Zeus had to dispose of him because everyone was getting better and overpopulation was threatening.

In Greece around Hippocrates's time—about 400 B.C.—hand healers were today's internists. They were called *cheirourgos* (the origin of our word "surgeon"), even though those *cheirourgos* used not scalpels but palm and fingers—particularly those fingers closest to the heart, the first and middle. Later the ring finger came into ascendance and was called the psychic or healing finger.

One of the earliest touch treatments consisted of a healer's lifting an illness out of a person and placing it in a tree or on an animal, like a goat—hence the term "scapegoat," someone who bears the blame for others. There were scapecats also—plenty of them. For eye problems, you would be well advised, in days of yore, to hold a black cat upside down and stroke your eye with its tail. That being not bad enough for the cat, many people held that throwing your chamber-pot contents over a cat would touch it with your ills and it would take off in a disgusting lurch for the hills, carrying your backache or scrofula with it.

And there were scapesnails, scapefoxes, and scapenewts. In some parts of England, rubbing the froth of a snail on an aching tooth would surely soothe it; ditto the Druid application of dried, calcined bodies of newts on the toothache area. Further, rubbing the dried lungs of a fox on one's congested chest would cure your catarrh nicely.

Off and on throughout early history, hands that were removed from bodies and pieces of hands were thought to have the magical power of touch healing. In Transylvania in the Middle Ages, the people who believed this developed the nasty habit of saving the hands or fingers of dead people as talismans against illness. Egyptian women were found in their tombs wearing fingers from Jewish dead as lucky charms.

As long as four centuries before the coming of Christ, ancient priests of healing traced arcane imagery on their patients' bodies, an act that is, even today, echoed in some Mediterranean countries, where Christian healers make the sign of the cross on a patient's head with a finger that has been saturated with spittle. And many of the alternative touch treatments currently in vogue in America still use these touched-on symbols; Reiki, for instance, which originated in Japan and emigrated to the West, requires practitioners to draw the ancient symbols on the patient's forehead.

Touch healing was universal in the early Christian church: Jesus touched crippled legs and gave them strength, touched blind eyes and gave sight, touched confused minds and gave them clarity. The Old and New Testament allude to touch healing, as, for example, this excerpt from Luke 4:40, "Now when the sun was setting, all they that had any sick with divers diseases brought them unto him; and he laid his hands on every one of them, and healed them."

Healing continued to flow from the fingers of the early Christians until the practice was dropped by the church during the seventeenth century, probably because of the church's decision to allow doctors to conduct autopsies on "bodies," while they maintained their control over "mind" and "soul"—thus beginning the split between body and mind/spirit. In fact, the liturgy of the "Laying-on of Hands and Anointing" was dropped from the *Book of Common Prayer* until, with the new edition in 1977, the laying-on liturgy came surging back in the Charismatic Movement of the Episcopal Church and reestablished the laying-on of hands as a force in United States religious practices.

One of the earliest Roman healers was Galen, who was born about A.D. 130 and, following Hippocrates's precepts, used gentle massage as a medical treatment. Although dissection of the human body was as yet a thing of the future, Galen was far advanced for his time in his knowledge of anatomy, and perhaps this was one of the reasons he understood the good of a rubdown.

Touch healers in the Middle Ages were called "chiothetists," and they were revered in the countryside, but gradually over the years they began to lose their clout. In 1423, the English Guild of Physicians denounced them as "quacks and empirics and knavish men and women." Less than a hundred years later, the English Parliament agreed; they let it be known that the only healers anyone of any intelligence ought to recognize were those who had trained at the College of Surgeons.

In the early days of France and England, the kings, who had undoubtedly begun to use the healing touch at the same time as the Christian church, took over. Now, it was called the "King's Touch" and the royals claimed to be the select possessors of the healing art. It was said that if a king touched you with his regal healing powers, you could say goodbye to many diseases, among them goiter and the unromantic-sounding scrofula, which, in 1058, was Edward the Confessor's specialty. He was simply terrific at scrofula.

During the Middle Ages, the kings continued their apparently effective touch healing, although it was increasingly being regarded with fear and suspicion as the work of devils or witches.

Still, none other than that very astute personage Samuel Johnson placed himself in a crowd to be touched by Queen Anne. Sadly, Johnson notes, he was not cured—but then, even chicken soup doesn't work every time.

King Henry VIII used to touch rings and give (more likely sell) them to epileptics.

It is interesting to note that touched money—that filthy lucre—was used for protection against disease in the reign of Charles II (1660–85). Charles treated more than 100,000 of his subjects by royal-touching a whole lot of coins, which were then greatly coveted by his people, who would buy the coins from the king. This benefited the subjects who reaped the advantages from the king's touch. The king's treasury, needless to say, must also have reaped advantage from his royal touch.

"Sacrament money," coins given in a church collection and later bought from the priest, was used to treat rheumatism and epilepsy in the seventeenth century. Its power was strengthened if it had been carried three times around the communion table before it left the church. The sacrament money was usually made into a chain or necklace by the buyer and was touched until it shone by its owner who yearned for the healing power that the priest's holy touch could transfer.

The royal touch was a different touch, depending on the royal who dispensed it. In medieval Denmark, the king could cure any childhood disease by touching the infant, while until just recently, in the Tonga Islands, the only disease the king could cure was scrofula—and he could only do that with a royal *foot* rather than a royal hand.

Once Charles II was so besieged by scrofulous Englishmen that over six hundred patients were crushed in a stampede to reach his royal fingers. In France, Louis XIV is said to have touched over sixteen hundred persons on one Easter Sunday.

Not until the late eighteenth century, however, did science turn its eye on the properties of touch healing. Franz Mesmer, a university-trained physician, decided that a "magnetic fluid" that emanated from the human body could be used to transfer healing energy from one person to another. This was the first scientific explanation for the effectiveness of the laying-on of hands and one that has shown further influence in many twentieth-century touch therapies. Mesmer conducted nicely scientific studies and much of Paris knocked at his door. He liked to dress himself in a long purple robe and carry a wand-type stick, an extension, actually, of his healing hand. He touched and healed an inordinate number of happy French citizens and was wined and dined by royalty.

However, by 1785, Mesmer's medical colleagues were miffed at his popularity and persuaded the king to set up a medical investigation commission. The commission quickly decided Mesmer was making outrageous claims, that his cures came only from exciting the imagination of his patients. But Franz Mesmer, although tried and convicted as a quack, remained convinced of his discovery of "magnetic fluid" or "animal magnetism." Forty-seven years later, another French commission decided Mesmer was, in fact, right but the new decision passed with little notice.

Meanwhile, other practices of Mesmer's—such as staring into the eyes of patients or having them stare at bright objects—were taken up for a while by doctors in France, Germany, and England. "Mesmerism," later renamed "hypnosis" (from the Greek word for "sleep"), found a sizable scientific following. These doctors, however, rejected Mesmer's original concept of "animal magnetism." And the idea of healing energy being transferred from one person to another was banished from science. "It remained," says the medical historian Richard Grossman, "only a religious or folkloric notion for over one hundred years."

Nor did hypnotism fare much better. After a burst of interest from physicians, it became more a music-hall trick than a healing art, until the neurologist Jean Martin Charcot began to revive it in nineteenth-century France. Today, however, both of Mesmer's notions are reviving and expanding with vigor.

Throughout the history of healing, many superstitions have grown up and some are still with us. Certain fingers of the hand, for instance, were thought to have magical powers. Grooms today invariably choose the third finger of the left hand on which to place a wedding ring because this finger has always been thought to lead to the heart. The forefinger of the right hand is sometimes called the poison finger, and even today some insist that it should never be used to apply medication (use the middle finger to be absolutely sure!).

The history of healing touch is varied and spans many generations, many centuries. Touch healing was lost and found and lost and currently is being found yet again. Dr. Lewis Thomas, author of *The Lives of a Cell* and *The Youngest Science,* put it best when he said that human touch was "the oldest and most effective act of doctors." He describes the days before science intervened with the stethoscope: "As a physician bent down and placed his ear over the chest and neck of his patient to listen for the noises of the heart and lungs, for diagnosis, it is hard to imagine a friendlier gesture, a more intimate signal of personal concern and affection than the close-bowed head affixed to the skin."

It is hard to imagine a more useful healing tool, as well.

In the following two chapters, I have divided the touch therapies into two, "The Touch of Energy" and "The Touch of Manipulation"; those techniques that rely primarily on energy balancing and those that depend primarily on direct hand pressure or massage. But, as I discuss later, my choices of which goes where are, to some extent, arbitrary. There's lots of cross-cutting between the two categories. Acupuncture, and certainly acupressure (described under "Energy"), could easily be called massage techniques; and Applied Kinesiology (included under "Manipulation") has an energy theory behind it.

Let's start with energy—a relative newcomer to modern Western medicine as we know it.

CHAPTER FOUR

HEALING:
The Touch of Energy

For thousands of years, highly developed Eastern cultures and many primitive tribes have viewed the movement of energy through the body—or the lack of it—as the very basis of health or illness. In the Eastern view, the body has meridians through which energy passes. In Sanskrit, this energy is called *prana;* in Chinese *Qi* or *chi;* in some African and Indian tribal dialects *mungo* or *mana.* In the West, we have neither a word for it nor a concept. Our nearest translation might be "vitality" or "vigor."

Eastern medical philosophy states that a healthy person has an abundant, rhythmical flow of *prana* or *Qi* and that an ill person has a distortion or a blockage of its flow somewhere in the body. In fact, having an unbalanced Qi is the Eastern *definition* of illness. The job of the Oriental physician is to manipulate key points along the meridians —as in acupuncture or Reiki—so that the Qi is again moving freely and abundantly through the body, thus healing any disturbed part. Also, since much of Eastern medicine is preventive, there are many practices an ordinary person can learn to keep the energy moving in order to buttress his day-to-day health. One of these is Tai Ch'i, which is simultaneously exercise and meditation, and is practiced daily by millions throughout China and in other countries.

The universal life energy itself comes from the total natural environment—the movement of moon and stars included—and is brought into the body primarily through the food one eats and the air one breathes. Thus, nutrition is important, as is living in tune with natural rhythms such as the waning and waxing of the moon, *and* breathing properly. That's why yogic exercises stress deep rhythmic breathing. Breath is the essence of life itself.

From time to time, this form of energy medicine has crept into Western countries, but it has never taken hold with any real vigor until

63

recently, when it seems to be entering more through the back than the front door: patients seek out Eastern therapies when their traditional doctors can come up with no remedies. The Eastern medicines were given a big push in the West in the seventies with the appearance of the humanistic psychologies and the emergence of the holistic health movement. But, for the most part, doctors themselves still remain highly skeptical, primarily because these treatments can't be explained by that holy of holies—the scientific method.

Well, okay. But what if the scientific paradigm itself turns out to be inadequate, simply not broad enough to explain these other older medicines? That's just what biological scientists are investigating right now and will be for some years to come. So we may be in for a rethinking of the whole scientific paradigm, just as we were in physics when it moved from Newtonian mechanics to quantum theory—a leap which brought the world into the atomic age.

In part, it's this shift in physics that's forcing the change in biology, for one of physics's new theories—known as the field model—states that energy fields are the fundamental units of all matter and all the environment. In this theory, a human being's energy extends slightly beyond what we perceive as his boundary—the skin—and he is interconnected through this energy with everything in the environment. If this is so, the energy fields within the body can be influenced—as in Therapeutic Touch (see next section)—by manipulating energy fields outside and immediately adjacent to the body.

In the meantime, something else is going on. A new field within scientific medicine itself—carrying the impossible name of psychoneuroimmunology—is beginning to confirm that our minds and our nervous systems are so tightly interconnected with our immune systems that our emotions, our beliefs, and our imaginings have an enormous effect on our ability to fend off diseases and cure those we do fall prey to. Such ideas are troublesome because the laws of modern science have been built heavily on the notion that mind-spirit and body are distinct and separate entities—a notion set in place by the seventeenth-century philosopher and scientist René Descartes. Descartes, it seems, is fast moving over the hill and, with him, part of current scientific theory.

Eastern medicine never fell into Cartesian dualism. Mind-body-spirit have always been one, which is why psychology, religion, and medicine are so closely wedded in the East.

However, certain Eastern energy therapies have gone out of style from time to time. During the Edo period in Japan (1603–1867), for instance, Shiatsu and other massage techniques fell into disuse because they were viewed as sybaritic pleasures rather than as therapies. In the mid-fifties, after a two-thousand-year gestation period Shiatsu was legally born again, sanctioned by the Japanese government.

But energy medicine is not just limited to Eastern philosophy and practices. Its notions have come from all over; indeed, it is so eclectic that everyone defines it somewhat differently. It can include things as diverse as acupuncture, diathermy, electrical stimulation, music therapy, ultrasound, infrared and ultraviolet light—to name a few.

A new research center in Phoenix, Arizona, the John E. Fetzer Energy Medicine Research Institute, started by a Michigan philanthropist who made his fortune in radio and television, is now investigating "various forms of vibratory energy, such as electric, magnetic and subtle" for possible new therapies. On its technical advisory board sit medical and academic scientists from the most prestigious universities and organizations in the country.

It currently divides its research into two areas: (1) "Energies which have been scientifically measured—'physical' energies such as electricity, magnetism, heat, sound, light, gravity and motion," and (2) "Energies which are only beginning to be scientifically measured—'subtle' energies, such as psycho-energetic, etheric, the energies involved in hope, faith, the will to live." Two of the purposes of the Fetzer Institute are to give energy medicine—especially the "subtle energies"—a scientific base, and to extend "health care beyond the traditional limits of allopathic (Western scientific) medicine, as the nature of man is redefined."

A wide variety of ailments can be treated successfully by energy medicine, according to the Fetzer people; among them are stress-related problems, pain of all sorts, brain and neurological disorders, paralysis, burns, lacerations, addictions, and learning problems. And there may be many more to come.

Almost all of the touch therapies can be related to energy medicine in some way or other. "Rolfing," for instance, which I have placed under "The Touch of Manipulation," also deals with the energy of gravity; and chiropractic (also under "The Touch of Manipulation") is built on the theory of there being a "universal intelligence" in all living matter which is carried throughout the body by electrical impulses in the nervous system. But, for the purposes of this book, I am limiting this "Touch of Energy" chapter to those therapies whose primary techniques modulate energy fields directly—either within or outside the body. Many of these touch systems have their theoretical base in Chinese medicine or have been influenced by it.

Since most energy therapies included here are gentle, most people can undertake them with no fear of ill effects. Some of them work better than others for certain people. But, of course, that's true for allopathic medicine as well.

I have indicated at the end of each section the conditions for which a particular therapy is best used. I sometimes list, as well, the conditions for which a particular therapy might be harmful—but *only* when such cautionary restrictions apply. Many of the treatments—such as Therapeutic Touch—produce no harmful effects no matter what condition you may have, and whether or not they may prove positively helpful. Under "Resources," (page 187), I give guidelines on how you can find the different practitioners in your locale.

Trust your own instincts on what you sense may be a good technique for you. We often know much more about the needs of our minds and bodies than we think we do.

THERAPEUTIC TOUCH

The *Ladies' Home Journal* brought Therapeutic Touch into my consciousness.

"Go write an article about touch healing," directed my editor at the *Journal*. "There's this nurse I've been hearing about—a real nurse who teaches about healing through touch—but it's not quite touch, either—oh, I don't know. Go find out. By the way—does anything hurt you? It would help."

It did. The curse of the writer—the perennially hunched-over-the-keys back was killing me. I had little faith in the laying-on of hands, but on the other hand could it hurt? I decided that it sounded like a timely assignment.

The "real" nurse turned out to be Dolores Krieger, a respected and beloved professor of nursing at New York University. Therapeutic Touch is a term she coined in the seventies for "an act of healing or helping that is akin to the ancient practice of laying-on of hands." But it isn't really laying-on of hands, because Therapeutic Touch rarely requires the practitioner to touch the patient directly but only to move her hands above the surface of the skin about two to four inches from the body. Contact, nevertheless, is made with the patient's energy "field."

In order to understand Therapeutic Touch, suspend for just a moment the ideas you think you know for sure—like the fact that you end at your skin. Imagine, instead, that there is a field of invisible energy that exudes from the body, a field of energy that is discernibly different around the areas that are unhealthy. For just a moment, test your own energy field, using Krieger's method from her book *Therapeutic Touch*, and see if you stop at your skin—or if your energy, as Krieger maintains, goes out beyond the obvious boundary into a "field" that, like an invisible aura, surrounds your entire body—head to foot.

See for Yourself

Sit comfortably with both feet on the ground and place your hands so that the palms face each other. Hold your arms away from your body. Now bring your palms as close together as you can without letting them touch. Now separate your palms by about two inches and slowly bring them back to their original position. Now bring them out about *four* inches and again return them to their original position. Repeat, but this time, bring them out to about *six* inches. Keep your motions consistent, slow and steady. Notice if you are beginning to feel a buildup of pressure between your hands. Is there any other sensation? Separate your palms again, about *eight* inches apart this time. As you bring them back, stop at about every two inches and experience the pressure field you have built up. Slowly, try to compress this field between your hands. You may sense a bouncy feeling. Continue experiencing this field between your hands and note what other characteristics of the field you may feel besides the pressure and the bounciness. Some people have sensations of heat, cold, tingling, or pulsations. Everyone is unique. You may feel nothing. You may feel something quite different from what has been described.

Dr. Krieger believes that it is possible to manipulate this invisible energy for healing. She believes—along with Eastern medical practitioners—that a healthy person has an abundant, regular flow of energy and that an ill person has a deficit or blockage somewhere in the body. With Therapeutic Touch, the healer smooths out the ruffled field of the ill person and channels new energy into the ill person's body. The new energy acts as a jog to the natural self-healing capacities that all of us possess. Her protocol is systematic and done in five distinct stages.

Dr. Krieger and her many students—known fondly as "Krieger's

Krazies"—have taught Therapeutic Touch to thousands of nurses, doctors, dentists, psychotherapists, ministers, and social workers in academic programs and continuing-education workshops across the United States and in thirty-six other countries including Australia, Ethiopia, Argentina and Zimbabwe.

All this I knew before I went to visit the "mother" of Therapeutic Touch to experience her method for myself. My general attitude was far from sanguine about what the results might be.

Dolores Krieger didn't resemble *my* mother at all; she has far too much zip, a sureness of attitude and confidence that is born of success, and—where she least resembles my mother—a trust in healing that can't be pulled from a bottle of pills. She is short, white-haired, sturdy, and practical—no more mystical than the old family doctor who made house calls. Dr. Krieger has never suggested that TT, as it is fondly called, is a substitute for chemical therapy or surgical procedures, although chemicals and surgery may sometimes not be needed if self-healing, prompted by TT, does the trick. She is the first to suggest that, if your appendix is in deep trouble and infecting the body, you get it out quickly and leave TT and its subsequent self-healing to the recovery stage.

And then there she was, providing me with the assurance that one needn't be a nurse or a specialist of any sort to learn how to "Therapeutic Touch" another. Anyone who has a fairly healthy body, a strong intent to help or heal ill people, and a mind able and willing to learn Therapeutic Touch can do it. "If I can do it, anyone can," said Krieger with a grin, "and once I found out what I could do, no one was safe. Still," she continued, "it's not just a matter of sticking out your hands and plunk . . . But what has to be done, a wife can learn, a mother or father can learn, a nurse can learn. It's a natural potential in all human beings and that potential can be taught.

"The first thing that has to happen is for the healer to 'center' herself—that is, concentrate on the task at hand and set aside all distractions and any concerns about how the therapy will turn out. It's a quieting process, a healing meditation."

This Professor Krieger proceeded to do. She sat silent for a moment, head down, absorbed in her inner self.

I, on the other hand, also sitting, hadn't stopped talking and had a hundred questions.

"Can you find it within yourself to quiet down for just one minute?" asked Krieger with a mischievous smile. "Notice what's happening inside yourself, as I'm doing. Rest your mind."

Next, she said, comes the "assessment." After the healer feels her mind and body are relaxed and concentrated, she runs her hands slowly along the patient's energy field to locate asymmetries—areas that generate sensations of heat, prickling, congestion, coolness, or thickness, for example. These presumably are the ill or weakened places. A novice cannot sense the rough areas in the field that surrounds the patient, but the experienced Therapeutic Touch practitioner comes to be sensitive to places that seem, well . . . just different.

Then, said Krieger, the healer must "unruffle" the field. This "relieves pressure and improves the energy flow."

Krieger stood up, beckoned me to do the same, and her hands moved rhythmically around me to brush away imbalances. Unerringly and without being told where my problem lay, her hands hovered over my back. I watched her in a mirror that faced both of us. What did it feel like to her, the imbalance in the field that surrounded my back? It was hard to say, she said—just something "too thick." One nurse, she said, described the sensation coming from a weakened area as "the force emanating from a fan placed on low."

The professor's hands continued unruffling. She swept them around my body, particularly in the back area, then shook them out, wiping away from my body whatever it was she was feeling.

I paid careful attention, knowing I would have to write about the sensation. In my head and chest a persistent drone seemed to quiet. In closing my mouth, I'd surely helped to mute the noises in my body, but something else was working, as well.

I felt, as I later jotted in my notes, as though "a giant rubber band —connected to her on one end and to me on the other—was being pulled out by Krieger and then released and let back to me. Out and back, out and back . . ." I might have been influenced by what I already knew about Therapeutic Touch. But I was working hard to retain my natural skepticism of things unexplainable. Was there a connection of sorts between Dolores Krieger and me—was there something passing back and forth between us? I didn't know. I still don't know—although I've thought long and hard about the new sensation.

The congestion seemed to have been loosened and the fourth step was taken by Krieger—that of "modulating" the energy. She used her palms, where the *chakras* or energy centers are said to be located, in an attempt to normalize, or balance, the area she correctly sensed was troubling me. Slowly, she moved her splayed, open-wide palms around my entire body, spending a whole lot of time on my back. She was, she later said, channeling energy through her body to my weakened area and thus helping to reestablish a rhythmical flow of my energy. Some nurses, said Krieger, used the power of imagination during this step, thinking hard of a cool color to relieve a hot spot, or a quiet, harmonious music passage to relieve a discordant spot.

Finally, Professor Krieger came to the fifth and final part of the Therapeutic Touch sensation. She gently stopped. Later, she explained that healers learn to sense when the weakened field of the ill person has had its energy rebalanced. To continue longer would overload the system.

Observing Therapeutic Touch in action is an odd experience. It looked funny—Krieger just moving her hands through the air around my body. Yet it has a very distinct procedure, and one that takes a great deal of practice to do well. The clue, says Krieger, is in training oneself to be sensitive "to very subtle clues—very subtle movements in the energy field." Like the acupuncturist learning to take his many pulses, it is a self-learning process—a process of tuning up one's sensitivities. An acupuncturist may take seven years to learn his pulses well. It doesn't take so long with Therapeutic Touch; the ability can be learned rather quickly.

Also, like acupuncture, Therapeutic Touch does not require any belief or knowledge on the part of the patient. The patient has nothing to do at all but sit, or stand, or lie there; and, says Krieger, a patient can be highly skeptical and it will still work. This right away sets it apart from faith healing, which depends on the belief of the recipient and the unique powers of a particular healer.

Professor Krieger was finished with the journalist. I tested my own responses from a very critical point of view. What exactly did I feel?

Great relaxation.

A general sense of well-being.

Perhaps the *tiniest* bit of relief from back pain.

That's all.

We talked some more, the professor and I, and then I went home, impressed with her verve, slightly disappointed in the results.

A week went by. At my typewriter, one day, I realized my back no longer hurt. Well, it felt the slightest bit tense from the day's work, but nothing like the way it had felt when I went for my Therapeutic Touch.

I am a convert. I think.

Dolores Krieger and others who practice this healing touch know full well that there are people who scoff.

"I welcome that," says the professor. "People have an absolute right to be skeptical of something that is quite far outside the mainstream of our culture. Better to be skeptical than embrace a theory with fake religious passion (there is *no* religion involved) and little real understanding."

How did it start, this practical professor's development of Therapeutic Touch?

In the sixties, Krieger began to hear of the work of Dora Kunz, a woman who Krieger says was born with a "unique ability to perceive subtle energies around living things." Working with scientists and medical doctors as well as lay people, Kunz became convinced that there is something about the universal field of energy that allows it to be channeled through one person to another and to be used to heal and relieve pain. She devoted herself to perfecting her own capacity to transfer energy and her abilities became, Krieger says, "like a fine instrument in her hands," which she could turn on and off at will. It was Kunz who first introduced Krieger to the possibilities of energy. And Kunz worked closely with Krieger as she developed her specialized protocol.

Among those people with whom Kunz worked was a layman, a gentleman named Oskar Estebany. Colonel Estebany had served in the Hungarian cavalry and one day his beloved horse had become quite ill. There was nothing to do but shoot it, said his superiors. Estebany was distraught. He spent the night in the stable, massaging and caressing the animal. In the morning the horse was fully recovered. Other cavalrymen soon brought their sick horses to Estebany, who would touch them with loving care, and often heal them.

The officer's fame spread. The children of the cavalrymen began to bring their pets, and the colonel healed them as well. No one, least of all Estebany, understood what was happening. Still, it worked.

Then one day a neighbor's child became desperately ill and the frantic father couldn't find a doctor. In desperation, he scooped up his son and took him to Estebany. No, said the colonel, his talent only worked with animals. The father begged. Estebany gave it a try. He touched and stroked the dying boy, and the child visibly improved. In a short time, he was well.

What happened? No one could say. Estebany continued to work on animals and humans. Eventually he retired. Determined to have people study his talent, he joined forces with a research group to which Dora Kunz belonged and in which the young nurse Dolores Krieger happened to be involved.

Today, Krieger remembers that the biggest surprise at first was the colonel himself. She'd heard of his touching successes and expected a prototype movieland healer with waving hands and a hypnotic glare. What she saw was a "well-built man with cheery blue eyes, a frequent smile and a deep air of commitment." He had no special mystical qualities at all, a fact that was deeply impressive.

Krieger worked with the research group, which included a medical doctor, Otelia Bengssten, M.D., Dora Kunz, Estebany, and a large sample of medically referred patients.

The healing session itself was a revelation. No mumbled incantations, no promises—just Estebany sitting quietly on a small stool in front of or behind the healee, reports Krieger. He gently put his hands where he felt they were needed and stroked softly. The patients seemed relaxed. After the session, most would report feeling better, but, to Krieger's disappointment, there were no immediate cures.

The real news came soon enough. In the weeks that followed, Krieger says she was "astonished by the numbers of medical reports that told of an amelioration or an actual disappearance of symptoms of illness." The patients had been suffering from a plethora of problems —everything from pancreatitis, brain tumors, emphysema, heart, kidney, and arthritic disorders to other weaknesses just as debilitating. And many were getting better, of that there was no doubt. Some of these were patients who had been written off by the medical profession.

Estebany's fame grew. Other scientists studied him. In 1965, Bernard Grad of McGill University completed some biochemical studies in which Estebany touched and "treated" some barley seeds and the flasks that held the water to nourish them. The purpose of the experiment was to see if Estebany could "transmit energy," as he phrased it, from his hands. The controls were strict, the results impressive. The barley seeds treated by the healer sprouted earlier, and the plants grew taller and contained more chlorophyll than the plants from seeds that were untouched by him.

Grad then conducted some experiments that involved using wounded mice, which Estebany touched. The mice treated in this fashion healed significantly faster than those that were not touched. Something that looked like magic but definitely wasn't was taking place.

More measurable experiments were conducted. In the mid-sixties a biochemist nun, Sr. Justa M. Smith, became interested in Grad's studies. As an enzymologist, she knew that if an energy change occurred in the mice that were touched by Estebany in Grad's experiment, it should be apparent at the measurement of the enzymatic level, for enzymes are essential to the basic metabolism of living organisms—and the energy inherent in them. Working with Estebany, she discovered that the test tubes of enzymes that he handled had chemical measurements that dramatically exceeded those he did not touch. The researcher concluded, along with the others, that Estebany's laying-on of hands contributed to the improvement or maintenance of good health.

Clearly the researchers were on to something. Krieger was deeply impressed and decided to do her postdoctorate work in the field—work that eventually led to her world-famous touch therapy and that gave the laying-on of hands a new structure and a new name—Therapeutic Touch.

Krieger met expected resistance as she worked. Despite the fact that touch has healed throughout history, Therapeutic Touch was virtually ignored by traditional medicine as a valuable tool. Health, lectured an undaunted Krieger, has to do with the balance, the ebb and flow of energy fields that every open system has—but few paid attention to her words. More determined than ever, she continued her own studies.

Controlled experiments were called for. She decided to work with the very matrix of humanity, its hemoglobin, the red, oxygen-carrying, indispensable substance in everyone's blood. In carefully monitored experiments at the Langley Porter Neuropsychiatric Institute at New York University, she demonstrated that hemoglobin levels and brain waves in patients changed during touch-healing sessions.

There were other experiments, other successes. Other nurses, Ph.D. candidates in particular, joined her in investigating different aspects of TT's effect. Gradually Krieger persuaded some medical colleagues, some institutions of learning, some hospitals to open their minds to the world beyond the traditional. Little by little, her work has become known in medical establishments.

Therapeutic Touch is still not a household word, and, say Krieger and other nurse-theorists, many more controlled experiments need to be made. Most of the nurse-experimenters believe that enough work has been done to prove that TT is highly effective in reducing pain—particularly in post-operative patients—and in lowering the anxiety of hospitalized patients and its accompanying physical tension. "If," says Therese Connell Meehan, R.N., Ph.D., of New York's University Hospital, "nurses can help decrease this anxiety and tension, they can facilitate their patients' own healing ability and that is tremendously important." With premature babies, this relaxation effect can be crucial. In many infant intensive-care units throughout the country, nurses use Therapeutic Touch to bring "agitated" babies back into balance so they don't exhaust themselves and jeopardize all their functions.

Many nurse-researchers—including Krieger—believe that Therapeutic Touch may prove to have even broader applications because of its ability to transfer energy and thus buttress a patient's immune system.

As with acupuncture, just *how* TT works has not yet been explained within the postulates of Western scientific medicine. We know acupuncture works—at least in specific areas such as anesthesia. "And," says Krieger, "we know Therapeutic Touch works—and therein lies its value. But we still need to prove *how* it works."

There are theories. One says that touching the energy field may stimulate our bodies' peripheral nerves, which, in turn, causes the central nervous system to release chemical endorphins—the body's natural painkillers. Another suggests that the energy passing from healer to patient during Therapeutic Touch is electromagnetic and, indeed, all people as well as plants and animals have electromagnetic energy emanating from them. But, at this time, most scientists see no way a human being can affect the electromagnetic processes in another; the fields are too tiny.

A third theory is that the results of this controversial therapy are due to the placebo effect, which occurs when both doctors and their patients believe so strongly in the treatment that the patients not only feel better but actually get better. However, as I noted earlier, Krieger and other nurse-experimenters record that Therapeutic Touch works well even when the patient is highly skeptical, and also works with premature infants, whose belief systems are not yet in place. Nevertheless, Krieger says that, even if this latter theory should turn out to be correct, between 50 and 80 percent of illnesses fall in this category, and placebos have been known to help over 30 percent of them.

Whatever the theories, Therapeutic Touch is steadily gaining new respect and wider acceptance. It deserves further study and greater application as a healing aid, say many medical researchers, because it *does* work.

At this stage of the game, when its benefits are observable but an explanation of how and why they happen is still lacking, the doctor of a thirty-six-year-old patient has perhaps the best attitude.

In 1975, Claire Cantrell, a nurse at St. Mary's Hospital in Philadelphia, had a brain tumor removed. She looked forward to a lifetime of drugs like Tegretol to control her persistent seizures. Then one day a friend asked her to be a guinea pig. The friend was learning Therapeutic Touch, and she needed someone on whom to practice.

Gradually, says Cantrell, after a few treatments the pain and seizures disappeared. She stopped taking medication. Her doctor was astounded. When asked how he felt about the treatment, Claire remembers with a grin, he said, "I don't care what you do as long as it works."

A reasonable response.

One point remains certain. Further investigation of Therapeutic Touch is necessary if it is to be incorporated into the household-word category. Demystifying the process will bring it into the public domain, even though it may be difficult to investigate a practice that relies on intent and invisible fields of energy.

New methods of study may have to be found to legitimize what thousands have already experienced. It won't be easy. It's hard to measure the unmeasurable, despite the fervent wish of touch practitioners that someone may find a "scientific" way.

Therapeutic Touch Is Best Used For

- Illnesses that are stress related. Since it helps a patient to relax, his pain and discomfort can be greatly diminished—often, also, his need for medication.
- Alleviating the pain from a multitude of problems, including headaches, rheumatoid arthritis, muscle spasms, stiffness and soreness of neck, back, or shoulders, postoperative wounds.
- Reducing stress and facilitating the growth and development of premature infants. Soothing irritable or sleepless babies.
- Relaxation before the administration of anesthesia.
- Relaxation prior to insertion of cardiac pacemakers.
- Relaxation for apprehensive dental patients.

For how to find Therapeutic Touch practitioners, see "Resources," page 189.

ACUPUNCTURE AND ACUPRESSURE

There are many stories surrounding the origins of acupuncture, but the one I like best I heard from Richard Grossman, who is the author of an extraordinary new book, *The Other Medicines.* It concerns some Chinese warriors in a long-forgotten battle. On the day of one particular skirmish, they had bad news and then good news. The bad news was that they lost the skirmish, and probably the war. The good news was that, as they found themselves at day's end, arrowheads stuck into various parts of their bodies, their arthritis had disappeared. Also, their tendonitis, bursitis, and a host of other afflictions that were plaguing the battalion. When the wisest of the Taoist sages back home heard of the happy coincidence of the injury cures, so the legend goes, they began to spend extended periods of time and meditation in an attempt to discover just what had happened.

The story sounds logical, even if it is apocryphal. There must have been *something* that prompted millions of Chinese to stick needles into themselves. I mean, you just don't do that unless you have a very good reason.

The fact is, the technique of inserting very fine metal needles into the skin to treat ailments and relieve pain has been around for a long time. There are 360 traditional body points in which acupuncture needles are inserted, although most practitioners today use only about 100 to 150 of those spelled out in the ancient texts. People opt for acupuncture to treat ailments as varied as arthritis, hypertension, ulcers, alcoholism, and overweight. In America, in the last twenty years, practitioners, echoing the ancient Chinese, use it with increasing frequency as aids in childbirth and surgery. Many prefer it to conventional anesthesia because of the low risk it carries; unlike chemical anesthesia, it will not depress breathing or lower blood pressure.

But, because Chinese medicine is basically preventive, many Americans, like the Chinese, are also beginning to use acupuncture for regular body checkups, and tune-ups to keep their energy flow balanced and their organs in shape.

Acupuncture is simply a form of acupressure. Both use the same pressure points and are built around the same theory. Acupuncture simply substitutes needles for the finger pressure of acupressure. No one knows for sure just exactly how either form works within biomedical theory but that they do work, when properly performed, few doubt.

Dr. David Eisenberg, a young Harvard-trained physician, was America's first medical exchange student to the People's Republic of China. He learned his Chinese medicine from the front row and describes it all in the ground-breaking book, *Encounters with Qi*.

Dr. Eisenberg points out that in Western medical schools the first professional instruments a student buys are a stethoscope, an ophthalmoscope and a blood-pressure cuff. The student of Chinese medical theory first purchases his acupuncture needles, which may vary in length from a fraction of an inch to seven inches and range in diameter from one seventeen-thousandth to one eighteen-thousandth of an inch. People who think of needles as being the size of hypodermic needles have to rethink the word "needle." Acupuncture needles are often no thicker than a human hair and bear little resemblance to large, hollow, injection-giving horrors. In China's earliest days of experimentation, needles were made of bone, bamboo, silver or gold, but today they're usually produced from copper or stainless steel.

Acupressure sites may also be stimulated with other forms of touch, using heat, electricity, cold, ultrasound, and even lasers.

The theory of acupressure and acupuncture is this: the Chinese discovered that certain spots on the body, when pressed, punctured, or burned, affect pain tolerance and the functioning of certain internal organs as well as the course that some illnesses will take. These spots can be close to each other or widely separated and still affect the same function. Pressing a point in an ear, for example, may do wonders for the calf. The points are connected by channels, called *ching* in Chinese, or "meridian" in English. The Chinese plotted the routes of these meridians, establishing twelve major ones, each associated with the twelve major organs. Each meridian has a point of entry, through which energy enters, and a point of exit, through which it leaves after moving along the meridian. Meridians are the "road maps" that enable one to locate specific acupuncture points. In addition to the major meridians, there is also a network of minor channels that are connected, so that the start of one is the termination of another.

The most important thing, and the most difficult for the Western culture to understand about acupressure and acupuncture, is the concept of Qi, which I described briefly in the introduction to this chapter. Qi is vital energy (often spelled *ch'i* and pronounced "chee"). The Chinese believe that Qi circulates throughout the body along the meridians, and controls the good health of blood, nerves, and vital organs. It must move freely if a person is to enjoy good health. However, Qi frequently becomes blocked; trauma, poor diet, stress—any number of bad factors—can stop the free flow. When this happens, the body is ripe for illness. Acupuncture is based on the theory that stimulating acupuncture points along the meridians rebalances and restores the flow of Qi.

"Where the needle is placed," says Richard Grossman, "is what determines whether energy flow is stimulated or dispersed. In each case the idea is to bring the flow into balance. When the balanced energy works as anaesthetic, it is affecting the transmission of pain through neurotransmitters. Acupuncture can either stimulate the flow of blocked energy or modify the excessive flow of energy."

Listen to Dr. Eisenberg describe the removal of a large brain tumor from the head of Mr. Lu, a fifty-eight-year-old Beijing University professor of history. The only anesthesia used was acupuncture needles applied with the additional stimulation of low-level electricity.

"Lu was not too enthusiastic about the idea," says Eisenberg. "He had only limited exposure to acupuncture and was not a devout proponent of traditional Chinese medicine. He was also understandably unnerved by the prospect of being totally awake and responsive during brain surgery. . . . The anesthesiologist selected six key points and inserted needles," two in the region of the eyebrows, two near the right temple, and two in the region of the left shin and ankle. Low-voltage electric stimulators sent controlled amperage through the needles to Lu's acupuncture sites.

"The anesthesiologist gave the go-ahead to begin," notes Dr. Eisenberg, "and the surgeons took up their scalpels. They made an incision along three sides of the rectangle outlined by a marking pen, and proceeded to lift a three-sided flap of full-thickness skin from Lu's skull. At the moment of incision, Lu failed to wince, grimace or give any hint of pain. He remarked that he was aware of the surgeons applying pressure to his skin, but that he experienced no discomfort. His pulse and blood pressure remained at their preoperative levels."

Dr. Eisenberg details every exhausting procedure of the operation. He says that throughout the entire procedure, more than four hours in length, Lu remained conscious, vital signs remained stable, and no anesthesia besides acupuncture was given. They conversed, says Eisenberg, the entire time Lu remained in surgery.

Now comes the dramatic part. After the surgery, Lu sat up, shook hands with everyone, including surgeons and observers, and walked out of the operating room, unassisted, thanking everyone, if not for a swell time, at least for an interesting experience. (His tumor, by the way, proved to be benign.)

Acupuncture is becoming increasingly well known in the West. It was regarded for many years as "fringe" medicine and was scoffed at by the medical profession. One British doctor, Felix Mann, studied acupuncture in France in the late fifties and attempted to teach it to small groups of doctors. He was met with general derision when he expressed the notion that sticking a needle in a person's foot could affect a kidney or a lung.

Then, public curiosity really began to grow when *New York Times* columnist James Reston came down with appendicitis as he was accompanying then-President Nixon to Beijing in 1971. He had little choice of anesthesia when he was rushed to a major hospital for an appendectomy. What they used for anesthesia in Beijing was acupuncture. Upon his return, Reston waxed lyrical, not only about the effect of acupuncture as anesthesia but about the pain-free recovery he achieved through its use. Reston blazed a trail and many who read his column decided to try the new/old method and attempt to do away with the complications that often occur when using more conventional anesthesia.

Reston was a convert and so were many others—including doctors —who came to experiment with it in the West. However, it does not seem to work for everyone. No one really knows whether this is due to genes or disbelief in the practice, or the doctor's lack of skill. Regarding disbelief, one fact is interesting: the Shanghai Acupuncture Research Unit, says Dr. Eisenberg, has used acupuncture on animals with great success, so it's highly unlikely that belief has much to do with it.

There were some who suggested that acupuncture worked by stimulating the production of endorphins, which are the proteins that affect mood and the perception of pain. That was an explanation the West could buy more easily than energy blockages. Endorphins are chemically similar to opium-derived narcotics and could very likely bring about an insensibility to pain without a loss of consciousness.

No one knew (or knows) for sure.

But the results are not to be argued. And the question might be asked: if the results for so many are so dramatic, how come more people don't ask to be punctured? After all, here in the West, we've been known to try a kaleidoscope of esoteric cures, ranging from sheep dung to copper bracelets to exotic plants. Well, for one thing, not everyone loves needles. Fear of them is so prevalent that one of the biggest problems in dentistry comes from phobics who will not allow novocaine injections—ever. Second, needles are not medically permitted with certain illnesses—such as certain kinds of tumors, severe hemorrhagic diseases, cardiac disturbances, and many other problems. Third, contagious diseases, among them AIDS, are thought to be spread through contaminated hypodermic needles, and some patients are very nervous about having *any* kind of needles employed on them, even though the risks in acupuncture are extremely minimal. Dr. David Bresler, the director of the Pain Treatment Center at UCLA Medical School, where acupuncture is used extensively, says that in over 3,000 instances—in which perhaps as many as 500,000 needles have been used—there has not been one case of nerve damage, infection, or organ damage of any kind. At Lincoln Medical and Mental Health Center in New York—

where acupuncture is used routinely for all sorts of addictions from overeating to drug abuse—some 130 patients a day have been treated over the last nine years with no side effects. Fourth, acupuncture, thousands of years old, is the new kid on the block in Western medicine. It may sound unscientific to many. Furthermore, not many practitioners understand or know how to use it.

There is no Western scientific proof, says Dr. David Eisenberg, that acupuncture points or meridians exist as physical entities. There is no consensus as to the best way to induce acupuncture stimulation, and possibilities include, besides needles, laser beams, sonar rays, and injections of water. And finally, says the doctor, most Western-style clinicians remain unconvinced, despite the evidence, that it does more than alter pain for a select group of patients. In fact, because of the lack of study and experimentation, as with many other touch techniques, including Therapeutic Touch, the practice remains in limbo in much of the West.

Perhaps another reason why acupuncture is not more widely utilized is that it's extremely difficult to learn. Dr. Eisenberg points out that inserting a hair-thin needle into "tough and resistant human flesh" is difficult. Students practice on pin cushions and layers of newspapers before they can manage to insert the needle quickly and smoothly into a human site. Seven or more years of training may be required to acquire the art.

What does an acupuncture session look and feel like? As a first step, the acupuncturist always takes the patient's pulse to help diagnose the illness. This is a very subtle, complex art (unlike pulse taking for general checkups) and requires much training and experience to do accurately. After the illness is diagnosed and when the first needle is inserted, most people feel only the slightest of pinprick sensations. There is almost never any bleeding during the entire process.

Richard Grossman likens the feel of the needle's insertion to the intensity of a mosquito bite, lasting for two or three seconds. "This sharp feeling," he says, "was quickly succeeded by a neutral sensation that I can only call highly focused pressure." Grossman was receiving acupuncture to mute the effects of an attack of sinusitis, which was so severe that his impeded breathing interfered with sleep. His acupuncturist, he says, was able to manipulate the energy flow of the meridians so that the congestion in his nose was greatly relieved, and he experienced a "direct local flow of the mucus that had accumulated" in sinus and nose.

Some patients report, instead of Grossman's "focused pressure," a sense of tingling or heaviness, and most say that the treatment is not painful at all. Many report feeling relaxed as well as relieved, even in some cases lightheaded or euphoric. Each treatment takes on average no longer than twenty or thirty minutes.

One evening I went to observe the patients at a family acupuncture clinic operated under the auspices of Montefiore Hospital in New York under Richard Grossman's direction. There Hispanic, black, white, and Oriental patients sat, some in a sleepy state, some chatting happily with others, all with needles sticking out of their legs, arms, ears. Some of the needles were applied directly to the skin surfaces. Others were placed through stockings or socks. Ailments were varied; some came to relieve arthritis, others stress, others to lose weight. A young woman doctor was cheerfully explaining and, as cheerfully, listening to complaints.

In America, acupuncture is regulated by each state and laws vary throughout the country. In most states, acupuncture can be performed either by a licensed physician or by someone supervised by a physician. In a few states, licensed acupuncturists, having undergone rigorous training, are also recognized as autonomous therapists.

"I can't believe I suffered so long before I found acupuncture," said an old woman, who happily demonstrated an uplifted arm—an act she says was impossible before treatment.

"I don't take sleeping pills anymore, or tranquilizers, and I don't do pot anymore to relax," said a young heavy-set woman. "I know I'll lose weight soon, also."

Grossman notes finally that touch is at the bottom of it all. Touch is a language, he says. It can nourish and it can also destroy. One of the most significant benefits of acupuncture comes not from needles or doctors, but simply from touch—the touch that gives well-being.

Some Chinese physicians, says Grossman, believe that, when the meridians and points were first mapped, practitioners used *only* their fingers or acupressure to manipulate the Qi and needles were introduced later to intensify the benefits. "In any case," says Grossman, "the use of fingers to manipulate the energy flow and relieve symptoms has endured through the centuries, and acupressure is a technique available to patient and physician alike for the temporary relief of distress, ranging from headaches and muscle tension to arthritic pain and menstrual cramps." Other practitioners agree, saying, in effect, that the needles are for the therapist—to give him concrete evidence that he's doing something real; the touch, the fingers, the human contact are the real treatment.

Acupuncture, says Dr. Adam Lewenberg, a New York State licensed acupuncturist, is "part of a new wave of alternative approaches to health care." He uses the treatment to attack many of the problems that baffle more traditional medicine, such as alcoholism, depression, migraine headaches, cigarette and drug addiction, as well as other modern plagues. It tunes up, balances, recharges, says Dr. Lewenberg, and it does these things dramatically.

Responsible practitioners of acupuncture and acupressure are quick to note that certain ailments are unsuitable for the treatment and these should be approached with more traditional medicine. Both American and Chinese practitioners are quite aware that some life-threatening illnesses or infections are better dealt with in other ways.

Leonard Lauder, the chief executive officer of Estée Lauder, Inc., is a case in point. While traveling with his wife, Evelyn, in the Orient, he became strangely ill. He had no idea of the source of his sudden weakness and pain. The two Americans visited a well-recommended Chinese doctor who touched the pressure points on Lauder's wrists with skilled fingers. "He played Leonard's arms like a violin," remembers Evelyn, "up and down, up and down, with consummate delicacy." Then, he told Lauder, "My medicine is not strong enough for you. Return to the West *immediately* and seek help from your doctors." The couple did just that, and Lauder was hospitalized with a near-fatal virus and told that his subsequent recovery was due to immediate action. Two more days and he might have died, thought his physician.

Acupuncture requires having an open mind, if the healer accepts you as a patient. Its premise involves history and assumptions that are alien to Western culture; one cannot read about meridians or energy channeling in a Harvard Medical School textbook. Still, even for those who question the theory of acupuncture, the treatment remains a viable option. You don't have to be a believer in Ch'i. It's virtually painless. And in five to ten treatments you'll know whether it's going to work for you or not. Clear-cut, no side effects.

Western medicine would do well to examine more thoroughly acupuncture's unexplored potential. Unimagined benefits may be the happy result, none of which would surprise those ancient Chinese warriors who 2,400 years ago found that when bad things (arrows in the body) happen to good people, serendipities may result.

Acupuncture or Acupressure Is Best For

· Control or elimination of pain (low back pain, shoulder pain, bursitis, tendonitis, pre- or post-surgical pain)
· Anesthesia—dental and surgical

· Muscle spasms
· Headaches
· Supplementary treatment to help regulate blood pressure
· Neuralgic conditions related to nerve injuries
· A broad range of addictive problems, including overeating, smoking, drug and alcohol abuse
· Depression and anxiety
· Promotion of general health. The Chinese go regularly to acupuncturists for a "tune-up" of energy. They use acupuncture as a preventive measure—a way to keep the body in balance.

Acupuncture Is Not Recommended as a Replacement For

· Medications used to treat infectious diseases
· Surgical (or medical) treatments for structural problems and abnormalities in the body, such as fractures, certain kinds of tumors, or misalignments

For information on how to locate an experienced acupuncture practitioner, see "Resources," page 190.

CHINESE MASSAGE

Chinese hand massage is yet another form of acupressure. As in acupuncture and Shiatsu, specific points along the meridians are manipulated to diagnose and heal. Chinese massage is a technique that uses a combination of rubbing *and* pressure (instead of just needles or just pressure) to prevent as well as heal illness.

Dr. David Eisenberg maintains that stimulation of the sites appears to correct body imbalances *before* they turn into problems that must be treated. Many adherents have regular Chinese massages even when they feel in the best of health. The doctor describes such a massage that he received from an old blind Taiwanese man.

His palms and fingertips scanned my body from head to toe . . . his hands were like electronic sensors. He began to push, pull and manipulate. Flesh ligaments and bones were putty in his hands. My muscles stretched, my vertebrae cracked and yet I felt no discomfort. Suddenly, Zhu found what he was after: a tender point, a "trigger point," one that if pressed with just the right force, sent a deep ache radiating to my extremities. "Spleen," he would mutter in Taiwanese, while manipulating the trigger point between his thumb and forefinger. "Kidney," he'd say about the next point.

As Zhu found each trigger point that located the organ responsible for Dr. Eisenberg's body imbalance, he'd apply pressure with two fingers, which produced, the doctor says, an odd sensation of fullness, warmth, and pressure. Then, the old man used the palm of his hand to massage the points.

Eisenberg was left in a state of bliss. He notes that the whole thing set him back one dollar and fifty cents, not a bad price for preventive medicine.

Massage recipients say that Chinese massage works dramatically but only temporarily. Benefits can be felt for only days and sometimes only hours. True devotees say they last much longer but only after sustained treatment over a period of years; it takes commitment to keep Qi flowing smoothly.

Again, in the West, few have tested Chinese massage either for the short or long run. Although it is available to those who search it out, Chinese massage is not as readily available as many of the other massage techniques.

Chinese Massage Works Best For

· Illness prevention
· Relaxation and reducing stress

See "Resources" on page 190 for suggestions on how to find its practitioners.

SHIATSU

Shiatsu is the Japanese version of acupressure. In Japanese, *shi* means finger and *atsu* means pressure. Like Chinese massage, the purpose of Shiatsu is to bring finger pressure onto certain trigger body points, which will then release energy blockage and allow the Qi to flow more evenly through the meridians. Shiatsu practitioners rely on prolonged and heavy pressure. Some say that Shiatsu on tired muscles causes reconversion of lactic acid into glycogen (a glucose-yielding substance in the body) and this eliminates fatigue and improper muscle contractions. Others say that the prolonged pressure works on the parasympathetic nervous system and tends to retard circulation. This puts the recipient in a deep state of mind-body relaxation, which redistributes the vital body energy. Whatever it actually does—and again, there's a dearth of Western medical literature on the subject—this is what Shiatsu looks and feels like:

A Shiatsu patient lies on a hard massage table, or, more likely, an exercise mat or even a carpeted floor. The massage is most standardly applied through only the balls of the thumbs, with the practitioner using both thumbs at one time, but occasionally the index, middle, and ring fingers are also used for treatment. Sometimes, a Shiatsu healer will use her palms or even her elbows, and during a visit to the Osaka Health Spa in New York, white-uniformed masseuses walked up and down my back, massaging the traditional trigger points with their talented toes.

It didn't feel terrific—not in the standard sense, anyway. Since the practitioner has to apply her full weight to the specific points, the sensation is somewhere between pleasure and pain, with greater impression on the latter. A complete massage usually covers almost every surface of the patient, who is lying flat with eyes closed and mouth slightly open. The pressure is applied only when the patient breathes out and the body becomes softer and looser. Shiatsu can be used to manipulate just a few specific troublesome spots as well as for an overall body massage. Because many of the trigger points can be reached with your own hand, Shiatsu remains a fine technique for self-massage. The key is in the amount of pressure; firm, steady manipulations produce best results. How much pressure? Place a thumb on a bathroom scale and lean until the scale registers 12 to 16 pounds (24 to 32 pounds with both thumbs)—a good, general pressure for most of the body. A manual on Oriental massage will diagram the key pressure points.

Tokujiro Namikoshi is the man who, almost single-handedly, popularized Shiatsu in Japan. In the middle fifties, the Chinese acupressure theories—or at least variations thereof—became legally sanctioned in Japan. Namikoshi claims to have treated over 100,000 patients and has trained, he says, more than 20,000 practitioners at his school. His book, *Shiatsu: Japanese Finger Pressure Therapy,* outlines treatment for a variety of specific illnesses, including diarrhea, asthma, rheumatism, and whiplash injuries. It presents Shiatsu techniques to eliminate fatigue, lower blood pressure, diminish pain, as well as methods for increasing general health and preventing illness.

In America, Shiatsu has been adopted by many young, stress-filled yuppies, who have found ways to Americanize the treatment. The walking-on-the-back refinement, for instance, is American. Modern fitness mavens have adapted the techniques but they often neatly side-step the original points and principles. Bonnie Prudden, for example, has come up with her "Trigger Point Pressure" program. Prudden maintains that her approach is derivative, not copycat; instead of using the prescribed Chinese trigger points (ordained for thousands of years, it must be remembered), she has found new ones—trigger points that are "unique" with each person. The ancient seers must be shifting uncomfortably in their long sleep. The point, say purists, is that one shouldn't try to fool Mother Nature. In this case, Mother Nature consists of "the real thing," not an Americanized version. Consider other "real things"—Japanese Zen and judo, neither of which has even found a corresponding American word for translation. These "real things" are impossible to translate into another language because the concepts they represent are so quintessentially Japanese in spirit and culture. Shiatsu is such a concept: there is, one notes, no corresponding American word for the practice. Unless its practitioners remain true to Japanese intent, it isn't Shiatsu.

Again, like most legitimate touch healers, serious Shiatsu practitioners don't brag of miracles. The treatment, they say, is not a cure-all for life's chronic ills.

The key word is always, in the end, "legitimate." Untrained practitioners give Shiatsu, as well as the other touch therapies, a poor name. There are those who claim to be proficient in the art of Shiatsu who have never heard of ch'i, Qi, or even vital energy.

Ron Rosenbaum is a journalist for *Esquire* magazine. He likened Shiatsu to a pointillist painting, created by tiny dots of color, which together make a radiant image. Shiatsu creates its effect point by point, and each moment of thumb pressure on a pressure point adds up to a "connect-the-dots" portrait of a healthy and energetic body, said Rosenbaum. "By the time the portrait has been completed, the unblocked energy that flows along the body's meridians does more than connect the dots, it illuminates the entire body with a radiance, a glow whose whole is greater than the sum of its points."

Shiatsu Works Best For

- Tension relief
- Physical therapy for rehabilitation and general body strengthening

- Illness prevention
- Fatigue

Shiatsu Is Not Recommended For

- A long-lasting cure for chronic disease
- Infectious diseases
- Disorders of the heart, liver, kidneys, lungs
- Cancer or tumors
- People who have just eaten large meals or those who are very hungry

For information on how to find Shiatsu massage practitioners, see "Resources," page 190.

REIKI

On the night I went for my Reiki treatment from the traditional Reiki master, Ellen Sokolow-Molinari in New York, I stopped to read her door, through which I would soon enter, and the door of her next-door neighbor—both of which were plastered with messages. On the neighbor's door, a variety of KEEP OUT THIS MEANS YOU stickers were pasted. Reminders of threatening guard dogs and rugged marines within cautioned all against trespassing.

On the door to the rooms where Ellen gives her Reiki treatments were stickers that spoke of peace and love and health; they were valiant little messages, some written in Japanese, put there, one sensed, as a gentle response to the hostility that radiated from next door.

Reiki is a system of touch healing that works by channeling energy through one person to another. The energy, which heals and balances, is "drawn" by the healee, and the amount drawn is determined by the needs of the person receiving it. The personal energy of the healer is not drained because Reiki practitioners draw from the universal life energy in the atmosphere; they simply act as channels through which energy passes to the healee. This concept, as you will note, is very similar to the principles of Therapeutic Touch and almost all Oriental healing practices. The idea of channeling energy from a universal source is basic to Eastern thinking.

Reiki, which means "universal life force" in Japanese, comes to us from Japan but was born in the Buddhist scriptures, the sutras, and written first in the original language of Sanskrit at least 2,500 years ago.

At the turn of the century, Dr. Mikao Usui, president of a small Christian university in Kyoto, found the formula and the four Reiki symbols which were the description of how the Buddha healed. Dr. Usui's successor, Dr. Chujiro Hayashi, built a Reiki clinic in Kyoto and to this clinic came a young woman seeking help for serious physical ailments. Hawayo Takata was cured; she became the next Reiki Master and brought the healing art to the Western world. Essentially an oral tradition, Reiki always was and still is passed on today from teacher to student in an apprentice-like relationship.

Based on the premise that everyone, even the very youngest child, has an innate capacity to heal, Reiki doesn't require great faith or even great concentration. It is a gift—accessible to all—and anyone can learn to be a healer of others and, more important, of himself or herself as well. In fact, Reiki healers are often seen walking down streets with hands pressed to a hurting muscle or throat, channeling healing energy to themselves.

I lay on a couch, and around me stood Ellen and two apprentice Reiki healers, because the channeled energy would be that much more powerful if more than one healer "worked" on me. Ellen was trained to be an architect and, although she has since devoted her life to Reiki, it didn't have to be thus: one can follow any inclination or career and still be a Reiki healer. Adrienne Mason, one of Ellen's apprentices, is a professional macrobiotic chef, and Sandy Argas, the other apprentice, is a college counselor.

The healing began.

Ellen placed her hands on my forehead in a process called "attunement," in which she drew with her finger the four Reiki symbols on the "crown chakra"—the top of my head. She thus, as she phrased it, "poured in" the symbols in order to jog the universal memory—the deepest part of me, which instinctively knows how to heal myself and others. Her hands felt warm—much warmer than normal body heat. The Reiki symbols, each one having a particular purpose, would activate and direct energy throughout my body in very specific ways. They would release energy blocks so the healing could take place. It was like tuning a radio so the sound could come through clearly.

And then, each of the women gently placed her hands on twelve different sites of major organs and the endocrine system. They kept them there for about five minutes per spot.

Ellen cradled my head with her linked hands, supporting it from underneath. Adrienne's hands were placed on my abdomen. Sandy kept her hands on my legs and feet. At least, I think that's who touched what. My eyes felt like staying closed in a lovely aura of nurturing and love.

Nothing much else happened. I noticed that Sandy's hands became especially warm—in fact, they began to sweat—as she encircled my right ankle, which had, unbeknownst to her, recently been sprained. Why were her hands so warm? I asked. "You seem to be drawing more energy here," she answered.

Hmmm. I was impressed. The three women murmured quietly among themselves as they touched, but did not massage me. Ellen said I was to try to concentrate on what was happening in my body, and they chatted quietly to me about what they were doing. They were, they told me, also receiving energy as they worked on me, because Reiki is a mutually giving system, a healing act that has to be experienced to be understood.

The women held me very passively. There was no rubbing. I felt, somehow, deeply in touch with my core.

Finally, Sandy "brushed" me down. With a fluttering touch from the top of my head in a downward motion, she smoothed out the energy channels and adjusted the energy circulation. It felt nice. I sat up, strangely light-headed. Ellen cautioned me to dangle my feet before I stood. I was profoundly relaxed, there was no doubt.

Reiki is taught in two methods. One is a hands-on method in which the student learns to heal herself, then others, then teach, if she wishes. The second method is an "absent healing" approach, in which one learns to use the Reiki symbols to send energy and healing to those who may be far away. Both methods require only an interest, practice, and a desire to heal. Practitioners claim that Reiki has been used successfully to heal serious as well as minor ailments, rejuvenate spirits, deepen spiritual growth and self-development. Ellen Sokolow-Molinari, a softly gentle, wise-beyond-her-years young woman, says that for her "Reiki is coming home, uniting with myself."

There are Reiki groups all over the world, and although they hold to similar principles, leadership seems to be a question that disturbs many practitioners. In 1982, a group of Reiki Masters decided to form the Reiki Alliance, which acknowledges Phyllis Lei Furumoto, granddaughter of Hawayo Takata, as the Grand Master in honor of "her direct spiritual lineage."

Other Reiki groups, Ellen's for one, insist that there is no "Grand

Master" of Reiki and choosing one only encourages separation and mystification. Yet another cleavage occurred when a master Reiki teacher split off and taught a related but somewhat different form of Reiki healing, called MariEl.

If you insisted on scientific documentation, you wouldn't find much in Reiki history, despite the many who claim personal knowledge of extraordinary healing. Like many other touch therapies, science has just not gotten around to studying it effectively, and thus it probably won't be taken seriously by anyone who relies on scientific proof and numbers. Erratic and unpredictable, the energy that comes from Reiki has not yet been studied and isolated by physicists, which just means, to healers and those who have been healed, that the physicists haven't yet produced instruments smart enough to measure it. Sometimes it works, sometimes it doesn't. The more open you are to receiving the energy, the more likely it will be received, but you don't have to have faith to be healed. Reiki masters have healed animals and plants, which don't have faith.

I don't know what I believe. The day after my Reiki treatment, my ankle seemed tremendously stronger. Maybe it was chance.

A postscript: the evening I went for my Reiki treatment, I was accompanied by my husband, who *does* know what he believes, and it isn't in Reiki. Since he is a notorious skeptic, he had to be convinced by the healers that he should take a Reiki treatment for his tennis elbow. He accepted—good manners, not curiosity, prevailing. Ellen, Adrienne, and Sandy worked on him. At one point, Ellen said, "Larry, turn over please. Larry, turn over, so we can do your back. Larry . . . Larry."

Larry was profoundly, intensely asleep.

It had nothing to do with the relaxing effect of Reiki, he maintained later that evening. Neither did the fact that his elbow no longer hurt. It would have improved without the treatment. "Just coincidence," said my husband.

Reiki Is Best Used For

- Stress, headaches, backaches
- Osteoporosis, shingles, back and spinal ailments, menstrual irregularities, and other illnesses

For information on how to locate a Reiki practitioner, see "Resources," page 190.

POLARITY THERAPY

Introduced by Dr. Randolph Stone, Polarity Therapy is an amalga-mation of manipulation, massage, energy, and postural techniques, which—like Rolfing—employs sustained pressure with a knuckle, thumb, or elbow. Its purpose is to bring about a realignment of the entire body posture, which in turn will release energy blockages and balance the body's "five energy centers," about which I will speak more later on. As George Downing, author of *The Massage Book*, put it, "Polarity therapy looks like Rolfing but thinks like acupuncture because they both share the theory of energy flow."

Dr. Stone was an Austrian who became involved in ancient Eastern healing systems, and after he immigrated to the United States in the early 1900s, he combined his ideas with Western medical theory, then refined and defined his own approach toward removing the forces that block the life force or energy.

Stone divided the body into poles similar to a magnet; according to his theory, the top of the body and the right side are positively charged and the bottom of the body, the feet, and the left side are negatively charged. The polarity therapist heals by transferring positive and negative stimuli through his own body to the patient's by using his hand as an active battery. (Again, this is similar to Therapeutic Touch, Shiatsu, Reiki, etc. It differs from these in that one of its primary aims is to realign body posture through massage.) Since the right and left hands have opposite charges of positive and negative, they can be used as opposite poles of an energy circuit: when they are made to have contact with another's body, a "jumper" cable results, in which energy can pass from healer to healee. Polarity Therapy, then, involves an exchange between the electromagnetic fields people possess.

Hard to swallow? Then consider the inescapable fact that many researchers have successfully employed sensitive volt meters to record the electromagnetic fields between plants and animals as well as peo-ple. For instance, two researchers, Davis and Rawls, taking electrical measurement of the human body, discovered that the right palm does indeed have a positive charge, the left a negative. Right palm energy strengthens (because it's positive), while left palm energy reduces (it's negative). To a polarity therapist, this is translated into using the right palm to bolster energy and the left to sedate and relax.

What's more, the researchers determined that clasping the hands together produces a closed energy circuit, which may, speculates Dr.

Russ Rueger in his book *The Joy of Touch,* explain the habit of hand-clasping in prayer. Counterpoint to the closed circuit would be a rubbing of the palms together, which would increase the energy flow—something many healers are seen to do, without even quite knowing why in some cases.

Polarity manipulation is one of those therapies about which little has been written, let alone scientifically tested. As a result, practitioners often differ in the application of pressure, with strokes ranging from deep concentration of thumbs to light hand brush strokes.

Polarity therapists often start sessions by manipulating the five energy centers in the body as the patient lies on a mat or table. According to polarity therapists, these centers each have separate jobs; one governs the voice, hearing, and throat, another the respiratory and circulatory system, another the digestive, a fourth the pelvic and glandular secretions, and the fifth, the elimination area—the rectum and bladder. Various manipulations are performed that will balance these elements and release energy blockages, and these manipulations range from high and soothing, to deep and painful, to stretching. One is, it's comforting to note, never asked to receive treatment greater than he can endure.

Polarity Therapy also includes special diet and nutrition plans, beginning with a cleansing program lasting sometimes up to two weeks; this purifies the body and helps it to expel toxins. Fruits and vegetables, liquids, and "the liver flush"—a drink made from a mixture of lemon juice, olive oil, garlic, and ginger root—are standard. Finally, a vegetarian diet is prescribed to create harmony and balance in the body.

Polarity Therapy Is Best Used For

No specific symptom or ailment. By realigning the body's posture, it aims to balance the life energy flowing throughout the body so the body possesses its maximum preventative and healing powers.

For information on how to locate a Polarity practitioner, see "Resources," page 191.

REICHIAN MASSAGE

In the mind-body revolution, Wilhelm Reich was a pioneer, at least in one facet of it. Long before most others—in the early part of this century—Reich said that the body reflected one's state of mind and that one's state of mind reflected the physical condition of one's body.

If you are angry, frightened, or depressed, he reasoned, you "carry" your body in the depressed attitude. The body is always vulnerable to a threat from within and emotional disturbances reflect themselves in physical manifestations. If the body is permitted to continue breaking down in response to destructive inner feelings, muscles become permanently rigid, contract, or thicken. And the more an individual's body tightens, the worse his emotional state grows. It becomes a vicious circle, in which both emotional and physical patterns exacerbate each other and harden into one frozen, debilitating condition. Not a pretty picture.

We need, Reich decided, a new therapy that works on both body and mind simultaneously—a special kind of psychoanalysis that includes body work.

Reich began as a disciple of Freud in the psychoanalytic movement. He didn't stop there. Spanning some thirty years, his career was mercurial and his eclectic vision led him into discoveries in such disparate fields as political psychology, "sex-economy," biophysics, meteorology, new forms of mind-body psychotherapy, energy medicine, astrophysics—and, of course, psychoanalysis.

Off and on throughout Reich's life, his ideas brought him political harassment and sometimes very serious trouble. He died in 1957 in Lewisburg Penitentiary in Pennsylvania, after having served nine months for contempt for refusing to appear in court on a charge against him: he had inadvertently transported his orgone boxes across state lines and violated an obscure statute. This action involved him in an interminable legal wrangle with the FDA. In spite of all this trouble, Reich developed a coterie of followers and admirers who are still very much in evidence today. His work has seeped into our culture and has influenced psychotheory, in particular.

Reich split from Freud when he came to believe that neurosis and most physical disorders were caused by a blockage in energy flow. Emotions and sexual sensations, he said, exist in the body in the form of energy and, if you block feelings, you block energy. The muscles build up "body armor" to defend against emotions and impede the flow of energy along the body axis. Full sexual release, which helps maintain energy balance, is impossible or difficult when these defenses are frozen in place. The goal of Reichian therapy is to break down the "body armor," analyze the conflicts and feelings that arise, liberate blocked emotions, and help reestablish a gratifying sexuality.

First the physical body armor and then the psychoanalysis, said Reich. If you relieve muscle tension first, you will be assured a faster emotional recovery.

Reich associated each different emotion with a different region of the body, which is one reason why Reichian Massage is a very specialized and systematic technique. The body armor, Reich said, runs in horizontal bands across the eyes, mouth and jaw, diaphragm, abdomen, and pelvis. Each band is a contraction of muscular bundles and each is related to different ego defenses. (The segments of armor and the emotions associated with each are all thoroughly described in Reich's classic book, *Character Analysis.*)

Reichian body work moves from the top down—on the higher segments first and then, over a number of therapy sessions, to lower ones—because Reich believed that the difficult pelvic blocks should be the last to go.

A therapist follows his own perceptions of where his patient is blocked most heavily and will spend more time on problem areas. Usually, his massage is far from gentle—it's a forceful kneading: a pushing and pulling of the skin, often alternated with tough finger jabs. All this in the service of getting rid of those tight blocks! Occasionally, he uses lighter stroking or other forms of touch—it all depends on the patient, where he is in therapy, and his particular requirements. As in psychoanalysis, the trained instincts of the "healer" are very important.

In addition to the basic prod, there are a potpourri of other techniques for different parts of the body: working with eye movements and facial expressions to soften armoring and elicit emotion; using the gag reflex to loosen throat blocks and bring out rage and sadness caught there; using sound—screams, cries, and groans—and movements—kicking, pounding, and reaching out—to reestablish emotional expression and energy flow.

The result from all this pushing and pulling, kicking and crying? "Well," said a young New York science writer, "the whole thing sounds rather strange if you try to explain it to somebody who's never experienced it, but I can tell you it changed my life. My whole body is more relaxed and my mind too. If you want me to sum up what I got, I'd say: relief, insight and the capacity for greater joy in life."

After Reich developed his specialized therapy, he sought to understand more about the nature of the sexual or life energy he was trying to release. He came to believe that it was omnipresent—it was not only *the* force in living organisms, but the stuff of the cosmos as well. He called the energy "orgone"; conducted experiments to prove its existence as an objective reality; and invented the "orgone energy accumulator"—known colloquially as the "orgone box." This was a simple device designed to attract a higher charge of energy inside; a place where people could sit and benefit from energy's healing properties. That was the device that got him in so much trouble.

Modern medicine has simply ignored or laughed at much of Reich's clinical and experimental work rather than evaluating it critically. Much of it has been misconstrued or taken out of context. His ideas have been viewed by many as extremism, even lunacy. However, the Reichian Massage is still very much alive and has been expanded and developed further by several generations of Reichian followers. And today his ideas—though modified—are being given more credence.

His theories about energy—give or take the orgone box—no longer look quite so strange to Western eyes. They have much in common with "the universal energy" concepts of Eastern medicine, and Reich was among the first to integrate the Eastern ideas with Western systems. He brought Freudian concepts about the unconscious into the energy paradigm and then presented scientific evidence for the existence and the nature of energy.

Reichian Massage should only be attempted by trained Reichian therapists as part and parcel of a total mind-body program that includes "talk therapy" as well. The emotional energies released through body work can be powerful, and the presence and support of an experienced practitioner are necessary.

Reichian Therapies Are Best Used For

- People whose psychological difficulties go hand in hand with physical tension
- "Somatisizers"—people whose psychic conflicts get channeled into specific physical disorders or illnesses, ranging from headaches to cancer
- Sexual blocks or dysfunctions that require getting in touch with body sensations as well as partaking in verbal analysis
- People for whom verbal psychotherapy has been unable to break through emotional blocks

For information on locating Reichian therapists, see "Resources," page 191.

BIOENERGETICS

Another touch therapy sprang from Reich's vision. Psychiatrist Alexander Lowen, a student of Reich's, developed Bioenergetics, which has a theory and therapy similar to Reich's and has taken Reich's ideas further into the arena of alternative healing and psychotherapy.

In *The Language of the Body,* Lowen says that buried emotions, consciously or unconsciously repressed, create chronic muscular tension and a loss of vibrancy in the body. Repressed desires of the present and long-buried emotions from infancy and childhood have a powerful effect on adult physiology. Like Reich, Lowen designed a therapy that combines "talk therapy"—analysis of childhood experiences and exploration of dreams—with body work.

But Lowen departs from Reich on some theory and some practice. He doesn't go along with the theories of the orgone (or those orgone boxes), though he does believe in life energy, which he calls "bioenergy." In contrast to Reich, Lowen is apt to begin his treatments with the legs, because of his idea that a person's contact with the ground is crucial to proper energy flow and "centeredness." He prescribes a number of very difficult postures, designed to set off vibrations in the legs—a proof that energy is moving there.

Like Reich, Lowen encourages screaming, crying, kicking, and lashing out to break through the blocks to the deepest layers of emotion in order to free both psyche and soma. He also recommends a number of techniques for manipulating body muscles. A common one is for the therapist to apply pressure and palpations to tight muscles on the side of the neck. He encourages his patients to yell or cry if in pain—and they often do so, letting out rage or sadness that has been held in for years. Another pressure massage is designed to loosen a tight jaw—often eliciting a similar response. "For many patients," says Lowen, "it is a new experience to let their voices and actions express strong feeling."

Of great importance are the techniques designed to "open" the chest and promote deeper breathing and therefore greater energy flow. Lowen discusses many of these techniques in his book *Bioenergetics,* and another book, *A Way to Vibrant Health,* is completely devoted to describing the positions, stances, and manipulations used to free the flow of energy through the body.

Many of the postures a patient is asked to take are "stress" positions, purposely designed as awkward in order to break through the physical defenses. One common breathing exercise has the patient bending backward over a short stool that is made more comfortable with a rolled blanket on top. Her arms hang backward over and behind her head, and she is asked to remain serene, breathe deeply, and hold the position for a limited amount of time.

"I do not wish to give the impression," says Lowen, "that pain is an essential part of bioenergetic work. As Arthur Janov points out in the *Primal Scream,* pain is already in the patient. Crying and yelling are one way to release the pain. The pressure I apply to a tense muscle is not that painful in itself. It is minor compared to the tension in the muscle itself and would not be experienced as painful by a person whose muscles are relaxed."

Because Bioenergetics can differ from practitioner to practitioner and patient to patient, it is difficult to describe a typical session. A therapist can use any number of different massage techniques and ask his patient to assume a variety of "stress" positions. Like Reichian Massage, Bioenergetics requires highly trained skills and a deep sensitivity to the special needs of each individual patient.

Although Bioenergetic treatments sound dreadfully grueling, if not painful, people who have undergone the therapy—and there are quite a few—swear by it. It is quicker and much more effective than straight psychoanalysis, they claim, and does well what it sets out to do: by freeing up constricting emotions and blocked body energy, it brings the body and mind greater aliveness and life itself a greater wholeness.

Bioenergetics Is Best Used For

· People with chronic physical tensions—muscles spasms or tightness not relieved by exercise or regular massage
· Psychosomatic ailments related to physical tensions
· Individuals who feel out of touch with their bodies—"spaced out" a lot, or not "grounded"—who feel that blocked emotions may be the basis for their difficulties

For information on locating Bioenergetic therapists, see "Resources," page 191.

CHAPTER FIVE

HEALING:
The Touch of Manipulation

THE RUB AND THE CRACK: MANIPULATIVE MEDICINE

What has no calories, is cheaper than an analyst, relies totally on specialized touch, and is a tonic for tension, lumbago, headaches, stiffness, writer's cramp, charley horse, tennis elbow, constipation, motion sickness, the common cold, depression and, say certain experts, a whole lot more serious, more explicit illnesses? It's massage: the greatest thing to come down the pike since chicken soup. Throughout the centuries, when people have been struck with a pain, a cramp, or a sting, the instinctive move is to touch the hurting place and rub it.

The healing therapies that are based on manipulation are, by and large, one or another form of massage. Sometimes the concept behind them is energy flow and its proper balancing, but the principal technique is the rub.

Although there are many different kinds of massage, many have one thing in common, and that is a steady rhythm. As with primal dances, love-making, and beating cake batter, establishing a regular tempo seems to be an essential ingredient of the art. You can change the pressure in a massage and you can change the site or movements in a massage, but the healing touch of hands (my mother called them "golden hands") should always be imposed with a steady, reasoned pace. Erratic and jerky or disjointed touches make for frazzled nerves. Disciplined, healing touch, a well-thought-out massage, is a source of well-being.

Massage is as old as the world. Ancient civilizations, like Greece and Rome, regularly used massage as a method for pain relief and healing. Hippocrates, writing in the fifth century B.C., said, "The physician must be experienced in many things, but assuredly in rubbing."

Pliny, the Roman naturalist, was rubbed regularly to relieve his asthma, and Julius Caesar was pinched every day to alleviate various aches, pains, neuralgia, and headaches.

Massage fell into disfavor in the West during the Middle Ages, when people became contemptuous of the pleasures of the body. This lasted through the Victorian era. Then, in part owing to the efforts of Pehr Henrik Ling, a Swede who combined his familiarity with gymnastics and physiology with ancient massage techniques, the Swedish massage heralded its return. The massage *was* the message for millions. It brought relaxation to body and psyche. It brought freedom from aches and pains. It reduced stress, improved blood and oxygen circulation, increased muscle tone, and promoted the smooth functioning of the digestive, respiratory, and nervous systems.

The good news was that you didn't even have to hurt to enjoy the palliative effect of massage. Anyone who has ever seen an athlete in quintessential good health receiving a massage from his coach or masseur has seen bliss. Massage can prevent hurting. Athletes who used to limit their touching to on-field "high fives" and postgame rubdowns now use massage as a regular training tool.

Can anyone have a massage? It depends on which massage and whom you ask. If one opts for a Reiki treatment, for example (see p. 87), it would be difficult to see how it could possibly harm since the procedure relies on a gentle laying-on of hands and not on a rub. It's important to be your own judge and question techniques that might seem hazardous. Certainly, experts at Swedish massage recommend that those who suffer from osteoporosis, varicose veins, herniated disks, inflamed joints, tumors, certain cardiovascular conditions, like thrombosis or phlebitis, should refrain from massage. So, too, should anyone who has had a blood clot, a hernia, or an abdominal mass. At the very least, if you have any of these conditions, you should check with your medical caretaker to determine if any risk is involved.

For the rest of us—heaven awaits, or, for some techniques, a temporary purgatory which then releases us to a wider heaven.

As you will note, many of these massage methods have as a basic premise the reactivation of body energy. They could be—and often are —categorized under "energy medicine." I have chosen to include them in this section because their basic technique is manipulation.

What follows are massage techniques that rely, each one, on touch of one sort or another. Each has its own unique ingredient that sets it apart from the rest. And almost every one is rich with history and supporters.

As in chapter Four, I have listed at the end of each section what ailments a particular therapy is best used for—and, where appropriate, those conditions for which certain therapy is *definitely* contraindicated. Again, you'll find ways to trace the practitioners under "Resources." As you read, trust your own gut feelings about what may be best for you.

ROLFING (STRUCTURAL INTEGRATION)

Ida Rolf did not start out as a massage therapist but a biochemist in Switzerland, and her contribution to the healing touch was a system called Structural Integration or, as it came to be called, Rolfing.

A tent pole, Ida Rolf was fond of saying, needs its pegs and guy ropes to support it all around, and a body works on the same principle. In the forties, she began to teach a method of strengthening connective tissue and muscles with concerted manual pressure.

Gravity was the most important element in the force and development of the human body's energy, she believed. When the body was correctly aligned, the gravitational pull of the earth allowed the body's energy to flow easily and the body itself was best supported—as it was meant to be. However, attrition of the years, accidents, and even emotional pressures could create displacements of the body, making the whole structure unstable and painful. You should, believed Rolf, be able to draw a straight line through ear, shoulder, hip bone, knee, and the external shinbone of the ankle. But few of us are so gravitationally perfect. The human structure tends to sag. Muscles work not only when they are called upon, as is their natural function, but also have to be used all the time just to hold up the body structure. Bad news: they became weak, flaccid, they developed adhesions, they allowed energy to be lost. Gravity was no longer supporting the structure but stressing it.

What to do? Rolf developed a method of deep massage that would reshape the body's physical structure, bringing it back into alignment. One doesn't reshape bodies all that easily, as the reader might imagine. Restructuring the connective tissue and reorganizing muscle relationships could only occur with the application of extremely heavy pressure, deep kneading, and rubbing of the muscles during a series of ten one-hour manipulations, usually a week or so apart. The therapist might use a knuckle, an elbow, or all the knuckles in a fist and often one spot might be worked over for several seconds at a time.

It hurts.

What's more, Rolfing did not involve just physical restructuring, because Rolf believed that feelings, too, created changes in the physical structure, as in being "bent with trouble, bowed with grief, or tight with anger." Restructuring the body also involved encouraging the patient to release emotional trauma as her body was Rolfed. Many say that Rolfing has a soothing effect on personality and behavior comparable to the effect of practicing Yoga over a long period. Others report recalling childhood memories that resurface during treatment after many years of being buried. The technique offers, then, not only postural changes to recover physical health, but emotional changes in attitude, behavior, and actions.

I have not been Rolfed in the service of this book. It sounded too painful and required too much time. It is a long process. Still, recipients of the massage say the treatment can be extremely exhilarating as well as painful; the pain, they say, is tolerable because it only lasts for two or three seconds at a time and stops as soon as the Rolfer withdraws her hand. Many feel that the experience is exciting: they can sense energy being loosened throughout their body and feel long-cramped muscles release tension. In addition, many claim that their abilities are enhanced and their attitudes and behavior are changed for the better.

Before the therapist "works" the connective tissue in almost every body part, including the inside of the mouth, he takes "before" pictures. Then, following the series of treatments, he takes "after" photographs. The changes in the body posture are often visibly dramatic. Some people stand as much as two inches taller—and certainly straighter.

It is important to note one thing: the ultimate aim of Rolfing is not the ramrod-straight good posture for which your mother always told you to strive or that your drill sergeant commanded. That may strain muscles. Rather, the goal is a proper alignment, a normal relationship between earth and body that leads to greater energy, a lightness or "lift" of the body, as well as equivalently positive psychological changes.

Rolfing Is Best Used For

- Poor posture
- Muscle pain
- Harmony of body and mind. Rebalancing the physical body releases emotional pain and traumatic memories; the release acts as a catharsis and one is less bound by unconscious pain and conflicts.

For information on locating a Rolfing practitioner, see "Resources," page 191.

THE ALEXANDER TECHNIQUE

In the late nineteenth century lived an Australian Shakespearean actor and monologist whose voice was his meal ticket, but alas, F. Mathias Alexander suffered from inexplicable periods of voice loss. From doctor to doctor he went, and no medical man found a medical reason for his flyaway vocalization. Nothing if not a dedicated person, Mr. Alexander spent the better part of nine years checking himself out. With a three-way mirror, he observed himself from every angle, studying, studying, studying F. Mathias Alexander, to Ms. Alexander's perpetual boredom. But no matter—F. Mathias prevailed: he discovered that his voice loss was directly related to a way he had of holding his head that resulted in a backward and downward neck pressure.

Shakespeare was soon forgotten. Alexander spent the rest of his life developing a comprehensive philosophy about learning and function. It was modeled on his newfound concept for changing bad habits, posture, and movements to allow newer, far better physical function. If you could, through massage and verbal instruction, actually bring about a natural, upward lengthening of the spine, thought Alexander, you'd really have an improved human body—able to move and think, work and speak with splendidly enhanced quality.

The ex-Shakespearean actor agreed with the thesis of chiropractic and osteopathy, which held that the correct alignment of the spine was the core of good health, but he felt the poor alignment of the spine was not due to accident or trauma but poor habits—a misuse of the body's motions. Further, he believed that the intellect could be used to replace bad habits with good, and therefore there was no one set of exercises that would do it for everyone. Each individual must be "reconditioned" at his own speed and according to his own bodily needs.

This created a problem. You couldn't write it down.

Neither can you write down golf, Alexander would reply to critics who clamored for a manual, a list—*something* to follow. But words were always inadequate to describe Alexander's exercises. You couldn't learn his system from a book. You had to be there, feel it, sense it, copy it from the master himself. Not so good for posterity, but Alexander wasn't troubled.

He began to train his students to use touch in a specialized massage that would "free" the neck, for beginnings. Spinal energy should flow in an upward direction, he taught, so that all movement came from the head. The recipient of his massage would feel his head float forward and upward, as if pulled gently by a giant hook, and feel his back widened and lengthened. Bad habits would gradually be broken, and joints and muscles, long cramped by misuse and disuse, would be freed.

How? No one yet knows. Still, if not hordes, then quite a few thousands have been liberated from pain and inhibited movement when they worked with an "Alexander Technique" teacher. Many famous people, including Aldous Huxley, John Dewey, and the Nobel prize winner Jan Tinbergen (who, incidentally, devoted almost the whole of his Nobel acceptance speech to the technique) sang and still sing its praises.

Generally speaking, teachers of the Alexander Technique vary their treatments in unspecific and slightly different ways. Some ask you to sit, others instruct you to lie down. The teacher speaks and the patient pays "channeled-in" attention to her words as she verbally describes the correct positions in which she places a neck, a back, a leg. The patient concentrates on her tension spots as the teacher repeats verbal instructions along with her physical manipulations. Most patients require a series of lessons over weeks, months, or years—depending on the seriousness of the poor postural habits and the amount of "reconditioning" necessary.

In 1976, researcher Franklin Jones put something on paper—some proof. He used X-rays, photography, and electromyography to examine those treated by the Alexander Technique. People were analyzed using their bad old habits and then analyzed again as their movements were guided by an Alexander Technique teacher. No one could doubt the results: their later functions were vastly improved.

Still, the world remains largely in the dark about this particular touch therapy. Secrets, passed down from teacher to teacher, remain with them. No manuals exist. The exercises are designed for the unique needs of each individual rather than for the public's general use, largely because of Alexander's belief that we must think of working not just on our bodies but on our whole selves.

The Alexander Technique Is Best Used For

- Rehabilitation of injured muscles or joints
- People who seem to be just generally unwell, tired, fatigued, without knowing why
- People who seem to have chronic backache, headache, and muscle pains
- People with high blood pressure, spastic colon, trigeminal neuralgia and asthma, among others

For information on locating an Alexander Technique practitioner, see "Resources," page 192.

The Feldenkrais Technique

Feldenkrais fans are legion. Before his recent death, Moshe Feldenkrais numbered among his pupils people like violinist Yehudi Menuhin; former Israeli Prime Minister David Ben-Gurion; anthropologist Margaret Mead; theatrical director Peter Brooks; Director of the Neuropsychology Laboratories of Stanford University, Karl Pribram; and even Julius Erving ("Dr. J" of basketball fame).

Feldenkrais, a Russian-born Israeli scientist, was also a black-belt judo expert, an electrical and mechanical engineer, a mathematician, a researcher on the French atomic-bomb program, and, some say, a disconcertingly arrogant and impatient man.

While he was still in his thirties, Feldenkrais was dismayed to have a flare-up of an old soccer knee injury. The recommendation from his doctors? Surgery.

Feldenkrais would have none of it. Like Alexander, he already knew something about the possibility of breaking long-standing and detrimental movement patterns through a reeducation or reorganization of the brain. You could change hurtful body movements, established even in early childhood, with special exercises. Muscles, even those long in disuse, could be reprogrammed to stimulate different parts of the brain and increase its functioning as a whole. The entire body could be taught to work to its optimum capacity.

Feldenkrais worked on his knee, exercising, looking for alternatives to surgery—and he found them. Not only did he cure his knee, but he found his niche in history. Like Ida Rolf, Wilhelm Reich, F. Mathias

Alexander, and other touch healers, Feldenkrais knew that the emotional and nervous systems can be organized to heal the physical musculature and the individual can engage in new learning—no matter how ingrained his damaging habits are.

What we are talking about, says Joan Pfitzenmaier, a Feldenkrais practitioner as well as physical therapist and Associate Professor and Chairman of the Program of Physical Therapy at Downstate Medical Center of the State University of New York, is a "sensory re-education through a non-controlling, non-directive, very giving, subtle touch." The therapist gives suggestions to the patient through her hands—what kind of movement to try, what might work—and the patient decides if she will accept the suggestion. A person can know what's good for her physically, even if she can't know intellectually—and, when her body tells her the therapist's suggestion is on target, she can figure out what postures are right for her.

The basic aim of Feldenkrais, said its founder, is to improve posture through self-awareness of stance, gesture, and movement. The body reflects what goes on in the mind and vice versa.

"If the neuromuscular system of a widow is depressed," said Feldenkrais, "and if she is taught to change the pattern, she will no longer be depressed." Despite popular opinion, he continues, one's habits can be changed. With subtle repetition, certain exercises can open the entire nervous system to new possibilities.

Like the Alexander Technique, Feldenkrais is almost impossible to describe. You must learn it from a practitioner who has studied four years to become proficient.

Within the system are two modes of learning. One is manipulative, a one-to-one contact known as Functional Integration. The other is learned in a group lesson, which can be taught to classes of up to three hundred people; this is Awareness Through Movement.

Using Functional Integration, the custom-tailored program, the practitioner's hands guide the client, without words, to greater self-awareness. The touch therapy is extremely gentle. About thirty different body positions are utilized.

People who have problems from injuries, chronic pain, neuromuscular difficulties, and even learning disabilities are well served by the individual contact. Functional Integration has helped cerebral-palsied children, stroke victims, paraplegics, accident survivors, and others caught in maladaptive modes of moving and standing and sitting.

The other mode in the Feldenkrais approach, known as Awareness Through Movement, consists of exercises carried out in group workshops, and is best suited for musicians, athletes, dancers, and actors who make their livings from finesse of movement. In both modes, all learn how to discover untried muscle patterns that work to free them from pain and awkwardness. All learn how to discard unworkable habits.

Feldenkrais did not emphasize the separateness of Functional Integration and Awareness Through Movement other than to say they were convenient labels for doing the same thing in different ways. "I especially don't like it," he once said, "if the distinction is made that one is for 'sick' or 'brain-damaged' people and the other is for 'normal, healthy people.' Which of us, after all, is not brain-damaged in the sense that we allow many areas of our brains to atrophy through misuse or nonuse?"

In both groups, people learn to pay attention to their moves, to *differentiate* so that the details of self or surroundings can be better sensed.

Most lessons take place with the pupil lying down to minimize gravity's effect as well as to induce a more relaxed state. The pupil is asked to use only the smallest, most subtle movements like turning his head or lifting his arm which he must repeat again and again—if they are deemed more acceptable than his maladaptive movements.

Concentration is of the utmost importance. "The pupil enters an altered state of consciousness, after a while," says Feldenkrais practitioner Pfitzenmaier. "I can tell when I touch him, ever so lightly, by the way he moves or doesn't move, by the heaviness of his limbs, by his breathing, which is the best motion for him, which to change, which to stick with. Did I say lightness of touch? Amend that to lightness except when it is heavy. The first principle of Feldenkrais is that there are no principles graven in stone."

The main object, said Feldenkrais, is "not to receive training in what one knows, but to discover unknown reactions in oneself and thereby learn a better, more congenial way of acting." The always subtle, repetitive movements that characterize Feldenkrais's exercises can gradually be honed down, refined, until only the thought or image of movement is necessary to achieve its end.

And where are the teachers qualified to touch-heal in the Feldenkrais method? There are, at this writing, about three hundred in all the world. Only three groups were ever trained by the master, although there are groups now embarking on the long learning period.

The "routine miracles" (as Feldenkrais called them) keep occurring. The Feldenkrais teachers keep listening with their hands. The young woman with cerebral palsy learns to walk with greater grace.

The violinist with the wrist fracture is "completely better in six weeks." The stroke victim finds her balance. Touch evokes its cures.

Feldenkrais Is Best Used For

· Neuromuscular problems, injuries, strains, and sprains—especially of the back
· Professionals like actors, dancers, and athletes who need to sharpen their mobility and versatility
· Handicapped persons, paraplegics, stroke victims, cerebral-palsied children and adults, who wish to develop greater control and balance

For information on locating a Feldenkrais Technique practitioner, see "Resources," page 192.

OSTEOPATHY

Andrew Taylor Still was a practicing country doctor on the Union side during the Civil War. When his three children died during an outbreak of spinal meningitis, Dr. Still was distraught and disenchanted with the way medicine was currently being practiced. The spine was the ultimate source of good health, he thought, and when spinal vertebrae slipped out of position or an abnormality of bone structure existed, a whole host of health problems would result. What's more, believed Still, the malfunction of joints stopped the flow of blood at the point of the problem and undermined the blood's ability to produce chemicals needed to fight disease. Dr. Still spent time observing animals, which, he believed, instinctively realized that their health was dependent on good spinal alignment. Cats and dogs, upon awakening, stretched out their limbs, making sure their musculature was aligned and functioning before they began to use their bodies. Humans also stretched on awakening, but then, posited Still, they hunched themselves over in a myriad of back-straining activities. Animals knew better and never compromised their spines.

Ever since childhood Still had been fascinated by bones. He used to dig out old Indian graves and exhume the skeletons in order to study them, a practice that no doubt cost him some popularity among his little chums. One day, when he was about ten, he had what would today be called a whopper of a headache. He started to swing, but his head hurt too much, so he let the end of the swing rope down to about eight inches from the ground, threw a pillow over it, and used the rope as a swinging pillow. He lay stretched on the ground with his neck across the rope, and voilà! the headache disappeared. Whenever he felt another headache in the future, he "roped" his neck. Years later, he surmised that it was the manipulation of bones that led to the free flow of blood and that, in turn, gave him ease.

As a doctor, years later, Still gave up prescribing drugs and tried, instead, to promote healing through what millions of children have known as "knuckle cracking." He placed tension on many joints other than knuckles until he could hear a click or pop. His colleagues thought little of this unorthodoxy.

Thus, osteopathy was born. From the late nineteenth century to the present era, the science has undergone many changes. What hasn't changed much is medical hostility toward osteopathy and other forms of manipulative medicine, like chiropractic.

Today an osteopath, like a chiropractor, is concerned with the establishment and maintenance of the normal structural integrity of the body. Both practitioners treat certain diseases and abnormalities by manipulating muscles, bones, and joints. What chiropractic calls *subluxations* osteopathy calls *spinal lesions*. (A subluxation is "a condition in which one of the vertebrae has lost its juxtaposition with the one above or below to an extent less than a luxation which is a severe misalignment of the joint that occludes an opening, impinges on nerves and interferes with normal neurological functioning."—James Stephenson, *Journal of the American Institute of Homeopathy*, 1955.) With the onset of the holistic health movement, both have moved closer to general practice. Many osteopaths treat with medication—herbal or chemical—as well as by manipulation; and many address infectious diseases and lifestyle habits as well as structural problems. Their primary emphasis remains, however, on keeping the body in structural balance so that it can heal itself.

Why would a person choose an osteopath rather than an orthopedic surgeon or a chiropractor? Very often, he wouldn't. He's been to an internist who's suggested rest. No relief. He's been to the orthopedic surgeon who's suggested surgery. No way. He's been to the chiropractor. Not much help. His best friend swears by an osteopath. The friend says he went for a backache that was alleviated and, somehow, his stomach problems seem to have disappeared as well. The recommendation is irresistible.

Osteopathy does have a beneficial effect on disorders not directly related to the patient's immediate complaint, say devotees. An osteopath tends to treat the whole body of a patient more pointedly than an orthopedic surgeon, and this kind of approach is very appealing to those who tend toward natural or holistic healing. For instance: in order to diagnose and treat a writer with a chronic backache, an osteopath might well watch her work a typewriter to determine what muscular strains are occurring. The orthopedic surgeon she'd formerly visited seemed to be primarily concerned with the specific physical evidence of strain in the body. Some people are beginning to substitute osteopaths for their internists because they see that the osteopath not only performs manipulations but, more and more, he also prescribes medications and gives nutritional counseling. In many states, the licensing requirements for medical doctors and doctors of osteopathy are the same, which buttresses even more the latter's appeal: it's rather like having two doctors for the price of one.

The primary difference between an osteopath and a chiropractor is that the osteopath manipulates soft and connective tissue structures before she gets to the bone or osseous manipulation, while the chiropractor usually gets right to work on adjusting the segments of the spinal column. The osteopath spends most of her time on balancing the relationship of tendons, muscles, and ligaments to bones. Chiropractors use more X-ray than do osteopaths, who in turn use more surgery and drugs than do chiropractors.

Osteopathy Is Best Used For

- Those people who want a primary physician who has a holistic approach to bodily ailments
- Those people who sense that their own life habits—especially the way they use and move their muscles—may be part of the cause of their discomfort
- Those who are more comfortable with soft-tissue manipulation as a prefix to (and sometimes instead of) vertebrae adjustment
- Back pain
- Arthritis, tennis elbow, osteoarthritis, and related complaints

For ways of locating a good osteopath, see "Resources," page 192.

CHIROPRACTIC

The debate continues.

"I'm going to see my chiropractor today."

"Are you kidding? You're taking your life in your hands! My ortho-pedist says that anyone who lets a chiropractor treat her for a bad back is deranged."

"Oh, really? How come you've had two spinal fusions and your back still hurts? And how come all I need is one little 'spine-cracking' treat-ment and I'm walking on air? Well—how come?"

How come, indeed. Chiropractic is one of the best-known touch therapies and there are currently thirty thousand, give or take a few, practitioners in these United States. They must be doing something right if there's business enough for all of them.

The word *chiropractic* comes from two Greek words, *cheir* (which, as you will remember, is also the base word for surgeon), meaning "hand," and *praktikos* meaning "done by." Manipulation of the spine and joints, done by hands, is probably here to stay.

Spinal manipulation is used by naturopaths and osteopaths as well as chiropractors. The treatment is described in the writings of ancient Egyptian, Babylonian, Hindu, and Greek civilizations that existed many years before Christ. The father of medicine, Hippocrates, left detailed instructions for manipulation and spinal traction. And, in more recent eras, a group of lay "bonesetters" plied their craft in England and other parts of Europe in the early nineteenth century.

Then—a long silence in written documentation. The world either lived nicely without bonesetters or, more likely, just neglected to make records. It is probably true that myriads of spinal and pelvic disorders, low-back problems, and cervical, thoracic, and lumbar strains were quietly being healed as never-to-be-famous bonesetters quietly cracked spines.

In 1895, an Iowan named Daniel David Palmer surfaced. He wasn't much of a doctor of anything, having specialized, up till then, in healing techniques relying on mesmerism (an early version of hypnotherapy). It was Palmer's idea that a "universal intelligence" permeated all living matter, and this intelligence controlled the well-being, indeed the very survival, of man as well as other organisms. It was the nervous system, thought Palmer, that pointed the way to health or illness and, if nervous impulses were blocked, the body would respond with a variety of ills. What could cause such a blockage? Palmer called it a *subluxation.* What body part Palmer specifically had in mind was the spine and the thirty-one pairs of spinal nerves that traverse the spine through the openings in the vertebrae. There are thirty-three vertebrae in the spinal column and they are just as neatly arranged, one on top of the other. When a bone loses its juxtaposing position with the vertebra above or below, the misplaced vertebra can obstruct an opening or lean on a surrounding nerve and thus interfere with normal neurological functioning. Adjusting the vertebra back to its proper place was Palmer's way of "unblocking."

It was occluded nerve function, rather than life energy or Qi of the ancient seers, that Palmer was intent on releasing.

In September of 1895, D. D. Palmer overheard Harvey Lillard, his deaf janitor, musing that his hearing loss of seventeen years had begun with a feeling that something in his neck had "gone." Palmer examined his neck, found an odd lump, and therewith performed his first "adjustment." Think of the bravery of Lillard! Unable to hear "the racket of a wagon on the street or the ticking of a watch," he still must have felt extraordinary alarm when Palmer "cracked" his neck. Lillard was made to "hear as before" to his and probably Palmer's delighted surprise. The realignment of the janitor's vertebrae somehow jogged his hearing back to action.

Despite this victory, until very recently chiropractic was not often used for general healing but primarily for illnesses more directly attributable to spinal problems. Chiropractors were concerned, almost exclusively, with the physical examination, X-ray and laboratory procedures necessary to evaluate and treat subluxations of the spine. Today, the situation is changing. A chiropractor's education includes at least six years of college and study in areas as diverse as anatomy, bacteriology, physiology, pediatrics, geriatrics, spinal manipulation, X-ray, and nutrition. They must also, by law, serve a clinic internship before beginning a private practice.

Included in an office visit given by a good chiropractor will be a consideration of the patient's whole lifestyle, exercise, and dietary habits, and what nutritional supplements he may need. The theme of modern chiropractic is health for the whole person, and the prevention of subluxations is as important as their realignment when they occur. The most effective chiropractors prescribe exercises to strengthen muscles that help to hold the spine in alignment. Many patients opt for gentle as opposed to vigorous chiropractic care, although both are options of the science.

What specifically does a chiropractor look for when you enter his office? Examination of the tissues surrounding the vertebrae can detect poor muscle tone, which might indicate vertebrae misalignment. Tender and sensitive areas can show neurological involvement—the blockage of neural impulses. A look at the way a patient moves may tell the chiropractor something: changes in motion ranges and aberrant movements can indicate congenital abnormalities, pathological processes, or problems with the anatomy. X-rays, of course, are useful in locating subluxations. Muscle strength, extremity length measurements, and an analysis of posture—and even skin temperature—can all indicate problems in the spinal structure and autonomic nervous systems.

There has always been antipathy between conventional medical practitioners, who are fond of comparing chiropractors to faith healers and quacks, and the approximately eight million Americans who regularly make the trek to their trusty chiropractors either as their primary source of medical attention or as a little extra adjunct to their orthodox medical doctor. They have no problems finding chiropractors because the profession is licensed in all fifty states, and Medicare, Medicaid, and many union and private health-insurance plans, as well as federally funded vocational rehabilitation programs, now cover its services.

Chiropractors, with their hands-on touch contact and natural forms of treatment, are now becoming more popular to Americans, who are finding out that popping pills and submitting to surgery are not always the best treatments for an aching back. Indeed, even some traditional doctors are starting (albeit slowly) to say that there may be a place for spinal manipulation in a world where mothers, fathers, children—everyone, it seems, including traditional doctors—has a backache.

Chiropractors, too, as I have noted, have been broadening their base by paying close attention to diet, exercise, and lifestyle habits. They frequently refer their patients, when they don't have the knowledge themselves—and many of them do—to nutritionists, exercise specialists, and acupuncturists, as well as traditional physicians.

If your back hurts, and traditional orthopedics hasn't seemed to lay a finger on your aches and pains, should you choose an osteopath who will work on soft tissue and connective-tissue structures before he gets to the bone structures or should you go to a chiropractor for that comforting (albeit sometimes scary) "click and pop" as he gets right to the adjustment of your vertebrae? Or should you go to a chiropractor who knows Applied Kinesiology (see next section)? Or try first by yourself with Touch for Health (see page 118)? There are no easy answers. There are good practitioners in all of these specialties, and the best answer is to try out a number of them for yourself. Your relationship to the practitioner may be more important than the specific treatment he uses.

Changing the musculoskeletal structure with the hands, instead of with a knife or drug, is an adjunctive approach to health care, and there's a tale of a miraculous cure on every neighbor's lips. People defend their chiropractors in much the same intensely personal way they defend their gurus or psychiatrists or nutritionists. Still, research would be nice. Research would give chiropractic, as well as other touch therapies, credibility for much of the not-so-brave middle class. But research means overcoming orthodox medicine's inclination to ignore chiropractic for reasons of its own. Indeed, one such effort, a meeting in 1975—initiated by the National Institute of Neurologic and Communicative Disorders and Stroke (NINCDS) to bring doctors of medicine, osteopathy, and chiropractic together—ended in friction and accusations, with the conclusion that no conclusions could be derived from the acrimoniously tainted get-together.

While the doctors struggle to gain or retain turf, touch healing goes on.

Chiropractic Is Best Used For

· Those who prefer fewer surgical and pharmacological remedies and more hands-on, natural forms of therapy, involving manipulation of the spine and joints, nutrition, etc.
· Those who have problems with low back pain, slipped disks, sciatica
· Those who have hip, knee, muscular aches and pains
· Geriatric patients with sciatica, loss of joint mobility, lumbago, and other problems
· Sufferers with migraine headaches

For how to locate a good chiropractor, see "Resources," page 192.

APPLIED KINESIOLOGY

Think of your body and its muscles as a balancing act.

First make a mental image of a group of acrobats forming a pyramid. If one of the acrobats weakens, the others will strain to make adjustments to keep the pyramid erect. But everybody will be off balance. All members of the team need to strain but there will be one who strains more than the rest because he has to bear the largest brunt of his teammate's debility; he's the fellow on the opposite side from the weak one.

Big aggravation for the pyramid.

Enter ringmaster, Dr. George Goodheart, a chiropractor from Detroit. In the early sixties, he invented a method for previewing the body's balancing act. He called it Applied Kinesiology—from the word *kinesthesia* (the sensation of movement or strain in the body's muscles, tendons, or joints)—and defined it as the "science of muscle activation." Goodheart put a lot more emphasis on muscles than did traditional chiropractic but, at first, he viewed A.K. only as a diagnostic tool. Since then, he has taken it a lot further than that. He has, in effect, broadened the whole practice of chiropractic.

When a patient is in pain, chiropractors traditionally test for muscle spasm, muscle weakness, or muscle tightness in order to pinpoint subluxations—the partial dislocation of the joints that are the source of much misery. But Goodheart took traditional chiropractic one step further. It wasn't the muscle that *hurt* that was the problem, but rather its opposing weak muscle, which then forced the original muscle to tighten up or contract. The weak muscle might not even feel painful. Remember the acrobat who starts to hurt because the fellow opposite him has weakened?

Every muscle, said Goodheart, has its opposite in motion. When someone reaches his hand forward, one muscle contracts and its opposite number must release. When he brings his arm back the "reach-out" muscle releases and the other one contracts. Experts in the field are fond of saying that these biomechanics work like a swinging door held in place by two springs so it can swing either way. If the tension on each spring is equal, the door works efficiently because the system is in balance. But, if one spring weakens, the opposite one ties itself in knots and the door won't work properly, to everyone's intense annoyance. Tinkering, cursing, or oiling never helps; only replacing or strengthening the faulty spring restores the balance.

The human body needs the same balance, reasoned Goodheart. If one has a weak or tight muscle—say the muscle that allows one to bend at the waist—and that muscle hurts or is in obvious spasm, treating it alone will be of little help. One must first treat the probable cause of the spasm, which lies in its opposite number—the muscle that allows

one to bend back up from the waist—the invisible source of trouble—
and then if necessary (it isn't always), also treat the hurting muscle.

There is more.

An applied kinesiologist sees, according to Goodheart's teachings,
a human being as an equilateral triangle—the three sides of the triangle
representing the structural, chemical, and emotional components of
the body. Good health is present when all three components are in
harmony and balance. If certain chemicals are missing from a man's
diet, the emotional and structural-muscular sides of the triangle can also
be compromised. If his muscles are weak, causing structural imbalance,
his emotional and chemical stability falter as well. Every part counts
toward strengthening or weakening the other parts.

Trace the odyssey of just one hip muscle. Your hip hurts because
a muscle is knotted, so you naturally favor it and change the way you
stand. The way your weight rests on your feet changes. Your posture is
affected. Your back hurts. Your emotional state becomes testy. Your
mind is not concentrating on work but on hurt. Everything changes—
mood, attitudes, body language, stamina—all because of one weak muscle. Pop an aspirin or a tranquilizer. Temporary balm at best, with
possible side effects.

The simplest, most permanent, and safest way to deal with the
problem? Locate the hurting muscle. Strengthen its opposing muscle
—the weak one. Then provide the body with proper nutritional and
emotional supports.

What does touch have to do with Applied Kinesiology? Only everything. Dr. Goodheart first taught his followers to diagnose by probing
the body with educated fingers to locate weak or inhibited muscles.
Then he taught them to treat their patients with standard osteopathic
or chiropractic techniques, as well as with the additional muscle-
manipulation techniques of kinesiology, to restore the muscle strength
that is essential in keeping the bones in place and achieving good
posture.

Muscle tests are usually performed while the patient is lying down.
Therapists alternate between one side of the body and the other to
detect differences. As these differences are spotted, touch and pressure
on appropriate points are applied.

The results were heartening. The new diagnostic tool proved excel-
lent, the theory of the opposing muscles proved valid. Patients responded rapidly, far more rapidly than if just their sore muscles were
treated. Those with intractable frozen shoulders, chronic sciaticas, pal-
sies, and other problems with the musculoskeletal as well as other organ

systems began to see rapid and encouraging results.

Through Goodheart's methods, chiropractors and osteopaths using Applied Kinesiology have also found that they can test muscle response for possible allergies—especially food allergies. This is how they test:

A doctor finds a strong and healthy muscle—perhaps an abdominal muscle, perhaps a back muscle—on his patient. He then asks the patient to put the suspected bit of food in his mouth, chewing it but not swallowing it. In a few seconds, the doctor again presses on the muscle to see if its ability to withstand pressure has lessened since the patient chewed on the suspect food. If the muscle goes weak, the patient is advised to avoid the food—at least temporarily. In the same way, a patient is asked to eat whatever food the therapist guesses may be missing from his diet. Weak muscles are then retested to determine if they were strengthened with the addition of the new food.

Applied Kinesiology is becoming more and more popular with musicians, typists, and dancers—those folk who must keep their muscles in top condition or they face professional loss as well as health disruption. Leslie Tomkins, a young New York violist, is typical.

She was in pain. The underside of one forearm was weak and shaky and hurt miserably. Her shoulders were cramped and sore. She had burning in her left hand and couldn't get it to move well on the viola strings. But she had to keep playing the summer season at Tanglewood.

She saw an internist nearby. He suggested aspirin and not playing so much. Not possible. She saw an exercise consultant, who gave her a massage and some exercises to do. Result: temporary improvement for a day or two but then, the same sprains and pains returned. Back in New York, she went to an acupuncture clinic. Again, some relief, but it was limited to a few hours—at best, a day. She saw a chiropractor, "who used ultrasound and moved some bones." She felt sick afterward. Somewhere in the odyssey, she was diagnosed as having carpal tunnel syndrome—a form of tendonitis, common among musicians and typists, in which the muscles swell and rub against the bones. She was told it could lead, in her case, to nerve damage that would permanently impair the full functioning of her hand.

Then Leonard Bernstein suggested she see a talented young man in Brooklyn who was studying to be a chiropractor but who already knew Applied Kinesiology through his work as a Touch for Health instructor (see next section). After testing her muscles, he told her that her whole posture was off—"that my body sank into my hips so that many muscle systems weren't working properly and were putting pressure on nerves and other muscle systems." Over a period of weeks—

with three or four sessions each week—he helped her change her basic posture, which in turn helped reactivate poorly used muscle groups throughout the body. He released some muscle spasms and helped balance body systems that had been drained through misuse. He advised a bland diet and specific vitamin supplements for her adrenals, which had been overstressed; he helped her by means of acupressure systems for dealing with "performance stress" (fingertips on the frontal lobes of the forehead, forcing a concentration on the music rather than on a generalized fear of performance). Says Leslie, "He dealt with me as a total person and major changes occurred in things I used to think had nothing to do with muscles. I have a lot more energy; I actually read music better; my whole body functions better for bicycling and things like that; and my arm, neck and shoulder problems have completely disappeared. Now I just go for a tune-up every once in a while to be sure I'm staying balanced. . . . I'm a sceptic turned proselytizer."

The aim of Applied Kinesiology—as with chiropractic—is to restore the body's "innate intelligence" (including its energy flow) through using natural therapeutic approaches to remove any obstacles that may be blocking it. As chiropractic has grown more holistic in its approach —often including such things as acupressure, herbology, and nutritional counseling—the art of Applied Kinesiology has expanded greatly the possibilities for diagnosis and treatment and put chiropractors on the cutting edge of the natural health revolution.

D. D. Palmer, the father of chiropractic, would be pleased with this technique that relies on the muscles to reveal the secrets of the body's weaknesses. It was, after all, his original idea that the body holds an "innate intelligence" greater than the intelligence of each of its parts.

Not all chiropractors practice Applied Kinesiology, but many do. It's best to check, if you're looking for a chiropractor who uses it.

Applied Kinesiology Is Best Used For

- General preventive health care. Applied Kinesiology practitioners believe they can detect a dysfunction before it reaches a symptomatic level.
- Diagnosis and treatment of muscular or joint aches and pains
- Food sensitivities and nutritional advice
- Poor body posture and ailments that may result from it

To locate a practitioner of Applied Kinesiology, see "Resources," page 192.

TOUCH FOR HEALTH

No man is an island.

Correct, says John F. Thie, doctor of chiropractic, smiling, vital, committed to touching for health and well-being.

"We are social beings. We need each other. We need to touch in ways other than for punishment or sex. But now we have another reason to touch—to help each other. We can touch for health."

And that is precisely what more than a million people, at his direction, have been doing. Touch for Health is not a doctor-to-patient therapy; it's a packaged do-it-yourself program for balancing the body's energy with a safe, easy-to-use potpourri of techniques taken from chiropractic, Applied Kinesiology, acupressure, and nutrition. It's all pulled together in a self-explanatory manual, *Touch for Health,* and you can learn it by reading and practicing on willing friends or family or by attending a Touch for Health workshop in your area. After looking at the manual, I'd opt for the latter. The book is reasonably clear, but then I've never been very good at following book instructions for fixing anything—my typewriter and car included.

By studying the manual or going to workshops, says Thie, anyone "with loving hands" can learn how to use acupressure massage points, simple nutritional guidelines, and muscle-testing and muscle-strengthening routines—all aimed at putting the body in better balance. The result: improved posture, an increase in energy, and a decrease in any existing pain. You become, in effect, says Thie, your own "health practitioner." He does not preclude using traditional medicine; he merely suggests that you can take more responsibility for your own well-being and enhance your overall health.

John Thie's father was a chiropractor and a naturopath. He was fascinated with the whole body systems and spent years working on various diets and nutritional regimens that would work as adjuncts to traditional chiropractic. He called his methods A New Approach to Therapy—not exactly the snappiest title in the world, but still an honest and informed way of looking at the whole body rather than just its hurting parts. Young John was his greatest disciple. He married at nineteen, went to work for his dad at their small pharmaceutical company, where his job was to call on chiropractors and say, in essence, "Won't you try A New Approach to Therapy?" Most wouldn't. Eventually, John became a chiropractor and began to follow in his father's footsteps.

And then he heard of Dr. George Goodheart and Applied Kinesiology, the theory that held imbalances were not due to obvious muscle spasms but to the weakness of the opposing muscle. Thie incorporated Goodheart's diagnostic techniques in his practice and knew he'd hit on something good. He tried to persuade Goodheart to write a book for lay people, but when Goodheart refused, Thie decided to create his own program.

What's new about Thie's system? What's new is that he's developed a system simple enough for laymen to learn and he's pulled together a number of techniques into a unique synthesis of what he considers to be the best of ancient theory and the best of modern theory. He has also made a few additions and changes to ancient theory.

Oriental philosophy, for instance, says that energy flows along meridians in the body, which sometimes become blocked. Acupuncture or acupressure can unblock the channel and free the energy. Thie agrees but he calls the meridians "acupressure vessels," and he maintains that the energy is represented by "free-flowing, colorless, non-cellular liquid which may be partly activated by the heart." Furthermore, says Thie, meridians have been measured and mapped by modern electronic methods that definitely prove their existence. The acupuncture sites where energy blockages may be released are clearly explicit in traditional Oriental body maps. Thie says these are "electromagnetic in character and consist of small oval cells called bonham corpuscles which surround the capillaries in the skin, the blood vessels and the organs throughout the body."

At some five hundred different sites, says Thie, you can stimulate this "liquid" energy to flow more freely. A Chinese doctor tests for energy imbalances by "reading pulses"—a very subtle and difficult skill to learn. Applied Kinesiology's muscle testing is much simpler—anyone can learn to do it rather quickly, says Thie—and it detects the same imbalances.

After diagnosis, Touch for Health has five major methods for strengthening weak muscles and restoring energy flow. You can use any one or any combination of the techniques.

The first addresses the lymphatic system, and you are taught how to manipulate "neuro-lymphatic massage points" throughout the body to reactivate the free flow of the system—which, unlike blood, flows in only one direction. If, through testing, you have found a weak muscle, you massage the lymphatic massage points that, according to the Touch for Health chart, are specific to that muscle. Since the lymph system acts as a cleansing agent for the body as well as providing antibodies and carrying food supplies to all the cells, it is important to keep the lymph system functioning well.

The second method deals with the vascular system—and its pressure points are principally on the head, where light touch is used to stimulate blood flow to the muscle in question and to its related organ.

A third method addresses the ancient energy meridians, which you are taught to massage with the flat of your hand in the direction of energy flow, thus opening up the channels for an easier passage of energy. The fourth technique—pressing the acupressure points themselves—sends further energy to the weakened muscle. Occasionally, it's necessary to "wake up" a weak muscle so the energy balancing "gets through." You do this with the fifth method: "juggling" back and forth the two ends of the weak muscle. After stimulating the muscle directly, you can then continue to balance the energy by the other systems.

Extraordinary and sometimes immediate results occur from using these methods, says Dr. Thie. For instance: a woman, considerably underweight, complains that she is depressed and given to crying jags. Some of her major muscle groups test weak. By the use of Applied Kinesiology techniques, she is also tested for food allergies and nutritional deficiencies. A small morsel of fish is placed under her tongue and she is asked to chew it thoroughly. Almost immediately, her mood improves and the muscles that previously were weak now respond with greater strength. How can she have reacted so quickly to only a bit of fish? Dr. Thie maintains the chemical reactions of the food with the mouth's saliva trigger the brain "to call out all the necessary action from the rest of the body to process the food." The woman's diet, as it turned out, was deficient in organic iodine such as that found in seaweed and raw or lightly cooked fish. Her muscles responded, as muscles do, to even micro-doses of the food.

Another person tests out with weak muscles. Through a combination of the strengthening techniques described above, blood circulation to the muscle and nearby organ is increased, the lymphatic system is stimulated, and acupressure points for the muscle and related organ are massaged. Blood, lymph, and energy should now be flowing more freely; the weak muscle itself should be strengthened, and the whole body should be in better balance. Repeating this procedure over a number of days or weeks, says Thie, should do away with any plaguing aches or pains, improve posture and mood, and restore energy.

Poor postural balance—and therefore poor muscle balance—is a grossly neglected area in American health care, says the enthusiastic

Dr. Thie; just through simple touch manipulations, people can easily learn "to use health itself as a preventative for disease."

Touch for Health Is Best Used For

· A do-it-yourself health-maintenance system—especially for body stress and temporary emotional disequilibrium
· Alleviating muscular pain
· Detecting and rectifying poor posture, muscle imbalances, and thwarted energy flow
· Detecting food allergies and nutritional deficiencies

For how to find Touch for Health manuals and workshops, see "Resources," page 192.

SWEDISH MASSAGE

Don't knock it, ever. It is Elysian in its virtues.

The Swedish Institute in New York is the oldest massage school in the country and *they* say that a good massage will rev up circulation, benefit the muscular and nervous systems, cut down fatigue, increase endurance, speed up the elimination of body toxins, and produce the most delicious, sensual, pleasurable haze around the body that a body ever dreamed possible. There are many who believe that touch was meant for massage—the all-over, nonintrusive, gentling, kneading kind of massage that is known as Swedish massage. It is communication without words—in fact, if someone wants to talk to you while he's massaging, give that short shrift. Words have little to do with the gift that massage is. If you are able to relax and go with the flow of expert touch, you will enter a state that is both dreamy and alive at the same time. Massage is a healing art, not a sexual statement, although, being sensual, when practiced by lovers massage can surely arouse.

Thousands of years ago, before civilization as we know it, many of the good citizens of the Mediterranean countries enjoyed full body massages every morning, massages that would free their muscle kinks, roll out their tenseness, awaken their nerves, and allow them to welcome the day with sybaritic pleasure. Percussion, friction, and rolling vibration met expectant Mediterranean flesh.

Fine aromatic oils were stroked and kneaded, glided and tapped, cupped and slapped onto pliant and willing skin for a good two hours before coffee (or whatever) was served. Today, clever and civilized as we are by twentieth-century standards, we have replaced all this with a stinging five-minute shower. O foolish twentieth century!

A Swedish massage is best taken in the nude on a massage table or floor—but not a bed (it's too soft). Oil is a must, whether it's baby, vegetable, or musk, because the key word in a Swedish massage, says masseuse Marguerite Scala of the famed Elizabeth Arden Salon in New York, is "flowing." "If you get a smooth and flowing massage, you will know what it is to be totally relaxed in body and spirit," says Scala. "Hours or even days later when tension returns, since you know what the quintessential feeling of *relaxed* is, you can duplicate it in your head, give yourself a mental massage, and find that it works surprisingly well!"

A warm, darkened room is a boon to massage. Music is nice. Cold hands and long nails belonging to the masseur are not nice.

Kneading movements, says massage expert Gordon Inkeles, triple the blood supply to the muscles, boost oxygen consumption in the muscles by 10 to 15 percent, encourage waste to be squeezed out of tissues, and promote healing and health by enlivening the circulation. Inkeles tells of the experiment by physicians who, as long as one hundred years ago, were concerned with the effects of massage on waste elimination. They injected India ink into the muscles of two rabbits and allowed each to resume their "hyperactive" lives, with one exception: the injected leg of one rabbit was massaged regularly. A month later, the rabbits were killed and dissected. Ink had stained black the muscles around the injection site of the unmassaged rabbit but no traces of India ink could be found anywhere in the body of the rabbit whose leg was massaged. But forget the science, forget the experiments—*feel* the goodness flow into you:

The masseuse moves up and down the spine, balls of her thumbs gliding, gliding, strokes, strokes, strokes the shoulder muscles; she kneads the neck, slaps the buttocks, works the thighs, presses, presses, presses, gently squeezes, pushes, pulls, squiggles, molds. She moves at an even, confident rate. She presses and strokes, presses and strokes— she never breaks the touch contact, as she varies the pressure—and lets the massage *flow*. She never presses on bone but concentrates on the meaty area of the muscles.

A Swedish massage is a skin orgy.

A Swedish Massage Is Best Used For

· Aches, pains, illnesses that arise from overloads of stress (including headaches)
· Strained muscles
· Fatigue or insomnia

Swedish Massage Is NOT Recommended For

· People who have tumors and lumps of undefined origin, thrombosis, fractures
· Skin diseases that may be contagious or worsened by manipulation
· Pregnant women (around the abdominal area)

For clues on locating a good Swedish massage, see "Resources," page 193.

REFLEXOLOGY (OR ZONE THERAPY)

Imagine this. You are lying comfortably, quietly on a massage table in a dimly lit room. Bach is playing somewhere in the background. A rolled-up towel is tucked under your knees. A young, sweet-smelling woman covers you with the blanket of your youth; you recognize it from the satin binding you used to love to rub on your face. Now how in the world did she find it? You start to ask and she cautions you not to talk.

"It will be pleasanter if you concentrate on what's happening," she says, smiling. "Don't make conversation. Just relax."

Relieved of the burden of being social, you do just that.

The young woman applies lotion (which she's warmed) to your feet —the only part of your body that the blanket doesn't cover. It's the only part of your body that she's interested in, anyway.

Then, slowly, methodically, lovingly, gently, firmly (so it doesn't tickle at all), she kneads, strokes, rubs, presses every conceivable point of your feet. You never dreamed your feet had so many points. You never dreamed you cared a whit about your feet. But, it turns out, you do. You love your feet. Having them rubbed leaves you in a dreamy, drift-away state. All 26 bones, 56 ligaments and 38 muscles in each foot sigh.

When it's over, you feel a bit woozy from the intense pleasure of it all. And maybe something else, besides pleasure. You go home feeling extraordinarily refreshed.

Later on, you notice your back seems to have stopped hurting. But that's funny. She didn't touch your back.

Regis Philbin, the television personality, had the same thing happen. He'd been plagued by a terribly painful kidney stone and was scheduled for surgery within the week. The same young woman massaged his feet. The next morning, he passed the stone. She hadn't gone anywhere near his body. Just those feet.

What's going on here? What's going on is Reflexology and the young woman is Laura Norman, one of the best of the best in the trade, the first reflexologist to be selected for *Who's Who in American Women*.

Reflexology is said to be an ancient Chinese art, at least five thousand years old. In the tomb of Ankmahar, an Egyptian physician in the sixth dynasty about 2300 B.C. was found a drawing showing one man giving another a reflexology treatment. An inscription above the figure reads, "Don't hurt me. I shall act in such a way as to obtain your favor." The way, according to the tomb, appeared to be through a nice foot rub. Similar techniques have been observed in some tribal communities but hardly ever in Europe. Reflexology was brought to the attention of Americans in 1932 with the publication of a book by Eunice D. Ingham called, *Stories the Feet Can Tell*.

Reflexology is based on the principle that the feet (of all things) are a perfect reflecting mirror for every organ, muscle, and limb in the body. When touch stimulation is applied to a specific spot on the foot, energy is transmitted backward through a network of nerves. In short, what you massage on your feet makes a direct difference to another part of your body. When a reflexologist massages your ankles, you won't *feel* a thing in your midsection, but your menstrual cramps may disappear. And if Norman were to do her little "thumb walk," that forward, caterpillar motion of a thumb bent at the joint and moving steadily along on the pads of your toes, you might be vastly relieved to see your sinus troubles clear nicely.

"Reflexology is never used to diagnose," asserts Norman, "and does not claim to cure." Still, the scores of athletes, actors, pregnant and elderly people, business people and, yes, even a surgeon or two who flock to her offices claim cures that include migraine, ulcer, and backache relief, weight loss, ease with pregnancy and delivery, energy boosts, among others.

What actually happens, say reflexologists, is that when the 7,200 nerve endings in the feet are stimulated, they stream out messages through the body. Energy flows through unblocked passageways. "Crystals," which reflexologists say are waste deposits that build up in nerve endings and capillaries, are broken up so they can be flushed out of the body.

As Norman rolls her thumb along my heel, I feel a sensation of mashed corn flakes. "Feel that?" she asks. "Those are crystals. As the stress and tension leave, the crystals are also flushed out, the blood gets flowing, which increases circulation, and the body is able to heal itself more efficiently.

On top of all else, the treatment, which can last from thirty minutes to an hour, gets rid of "mental chatter," says the practitioner.

The feet, says Norman, are mini-maps of the body. Imagine the foot being divided by a "waistline." The pads of the toes represent, in a general way, head, eyes, sinus; the ball of the foot is the general area for the heart, lungs, and environs; the midsection of the foot is the stomach area; and the heel, the lower back.

"I was living in a nightmare of pain," says Sandra Thompson, a Long Island housewife referring to her damaged sciatic nerve, a long-standing condition that kept her bedridden and in traction for weeks at a time. Today, with reflexology, she is cortisone-free and relatively pain-free.

"Many people tend to dismiss us as quacks," says Laura Norman cheerfully. "But they change their minds."

Susan Berman thought she was dealing with a quack. Then the stockbroker says, "I was induced to try it. Could it hurt? No. Could it help? All I can say is that three months of regular Reflexology and I was rid of twenty years of constipation."

The best part of the whole thing is that you can learn to do it yourself since the feet are so accessible. Whether it cures or not, it surely relaxes. Reflexology is one of the least complicated, least mystical of the touches of well-being. Ask Susan Berman.

Hand Reflexology, I hear from devotees, is also wondrously relaxing. It works in much the same way as foot Reflexology—with specific points representing the various parts of the body. I haven't tried it yet. I'll hang in with feet a little longer.

Reflexology Is Best Used For

- Relaxation
- Symptoms of stress, headaches, backaches

- Arthritis
- Bedwetting
- Fever control
- Relief of normal pregnancy and childbirth problems (nausea, cramping, edema, labor pains)

For information on locating a Reflexology practitioner see "Resources," page 193.

MEDICAL MASSAGE AND HOSPITAL TOUCH

In the good old days, touch was the best potion a doctor could dispense. But, with the spectacular advent of modern bioscience, touch went completely out of fashion. Medical reports of the seventeenth and eighteenth centuries cited innumerable case histories of how massage helped cure a variety of disabilities. By the latter part of the nineteenth century, most of these reports disappeared from medical journals, and those that were reported were often discounted. Medicine was so busy with—and so enamored of—its amazing technological advances that it forgot all about touch, or eschewed it as trivial. For diagnosis, instruments like the stethoscope or more advanced machines like the electrocardiograph replaced the very personal eyes, ears, and fingers of the doctor. For treatment, touch was pushed aside in favor of surgery or drugs—especially when antibiotics arrived.

Then, following wonderful cures of infectious diseases, bioscience began to learn that pills and surgery were not always effective for preventing and healing many chronic conditions—like heart disease, cancer, or long-term pain syndromes. Pediatrics found out that no quantity of pills would help an infant or small child who was not being cuddled and stroked. Obstetricians began to see that, in the delivery room, touch not only enhanced a mother's experience of birth but contributed in a major way to the bonding process for mothers and fathers alike. So touch began to creep back into medical and nursing practice.

World War II sent home many soldiers with amputations and injured limbs, which further surgery could not ameliorate. Only rehabilitative massage could help, and in 1947 a new medical specialty

—physiatry—was created to address the special needs of the disabled. For other conditions, ancient Eastern touch therapies—and some new ones from the West—were given a boost by the humanistic psychology and the holistic health movements of the seventies—both of which stressed psychosomatic factors and preventive as well as curative techniques.

Today, although the healing potential of touch is far from being fully explored or exploited, hospitals are beginning to encourage their nurses and doctors to take training in touch therapies, and to be more aware of the use of simple, compassionate touch in their daily rounds. Most of today's administrators and their medical and nursing staffs have waked up to the fact that many of the new technologies, as helpful as they are, have obstructed much of the art of person-to-person healing. In one New Zealand medical school, touch is the first subject introduced to students when they begin their medical training; and clinical demonstrations of how and when to use touch are given throughout their first year.

Before I turn to examples of the uses of touch in hospitals, let's take a look at those physiatrists—who make use of touch perhaps more than any other of the medical professions.

You'll find most physiatrists in special centers for long-term rehabilitative medicine—like the International Center for the Disabled in New York or the famous Howard A. Rusk Institute of Rehabilitation Medicine, affiliated with New York University's School of Medicine.

Physiatrists treat anyone from an infant born with physical abnormalities to an eighty-year-old with complications from stroke or arthritis to an athlete with a spinal-cord injury. They analyze a patient's condition and prescribe whatever combination of massage, exercise, hydrotherapy, and/or other techniques or drugs are appropriate. Normally, they hand over the therapy itself to a trained physical therapist, but they are well aware of how to do it themselves. "Rehabilitative massage," says Dr. Milton Lowenthal, the past Medical Director of the International Center for the Disabled, "uses any combination of techniques from both Western and Eastern theory. It is a form of neural modulation—along with other forms like exercise and biofeedback—which reduces pain by helping the brain return to a normal balance, after pain has created an imbalance." A therapist at ICD says, "People who come to us have arthritis, strokes, neurological problems—even gunshot wounds. The medical massage helps break their pain cycles by influencing lymph, blood, muscular and respiratory systems."

Working with hard touch or soft, probing or palliative, stroke or push or tap, therapists daily ease pain and soothe spasms.

I saw a seventeen-year-old boy at the International Center for the Disabled—a boy who, in the prime of vigor, had dived off a board into a pool that held only two feet of water. He broke his neck, and the use of his legs was gone forever. But he could, theoretically, use crutches. "Theoretically" cut no ice. The boy hurt. He couldn't hold the crutches, let alone hold himself erect on them. Medical massage after massage after massage loosened his muscles, muted his pain, and the day I was there to visit he walked, strong arms propelling his body and useless legs. Massage had given him mobility. He was no longer sentenced to a rolling chair. It was extraordinary to see. The primordial form of healing—human touch—had accomplished what scientific technology could not. There could be no more powerful argument for the integration of touch with the more "sophisticated" medical repertoire.

To receive therapy in a center for rehabilitative medicine, a person is usually referred by his physician or hospital. At the very least, he must be seen by a physiatrist, who is able to prescribe the specific combination of treatments that will help best with his particular condition.

Now let's turn back to touch in hospitals. It is being revived in so many forms and as the result of so many studies and the experience of so many experts that I can only offer a pastiche of examples:

ITEM: Ada Rogers, R.N., a research associate in the Analgesic Studies Section of the Sloan-Kettering Institute for Cancer Research, says, "I've been working with cancer patients for thirty years, and for twenty of those years I've been involved in analgesic studies to see which painkillers work and which don't. There is no doubt in my mind that touching a patient is enormously helpful in relieving the pain of that patient. I hold patients' hands, I stroke their foreheads, I kiss, I hug, and I know it's the most important thing I do. Even the pain medication seems to work more effectively when I touch them before I medicate."

ITEM: Dr. Frederick J. Stoddard, Director of Psychiatry at the Shriners' Burn Institute and a member of the Department of Psychiatry at the Massachusetts General Hospital, says, "One of the most important aspects of rehabilitation is physical therapy, which is practically all touch. After an acute illness or severe physical trauma, touch inspires muscle action, promotes psychological healing, and is a major anti-depressant."

ITEM: A nurse in the Shriners' Burn Institute: "A child who is burned and cannot be touched normally is in deep trouble if his care-takers can't use ingenuity and provide holding and touching in different ways. To be held is to be comforted, to be *contained*. Sometimes a story or a game can provide a close facsimile of being touched and the child or adolescent receives an "inner" touch experience, so to speak. Without touch, recovery is doubtful."

ITEM: Dr. Mary Howell, an M.D. affiliated with Massachusetts General Hospital and Boston's Children's Hospital: "I use Therapeutic Touch to relieve pain and I also use it when dealing with connective-tissue injuries; I believe it promotes healing in ligaments, strains, and sprains. We've had a kind of selective blindness in this country about the connection between healing and touching . . . there's too much evidence around that it's real for me to ignore it in my practice. I don't substitute touch for drugs or surgery when they're appropriate, but it's definitely a therapy, just as X-rays are a therapy and just as chemicals are a therapy."

ITEM: Psychologists Sheryle J. Whitcher and Jeffrey D. Fisher at the University of Connecticut found that touching women patients prior to surgery brought fewer worries from those patients about complications, a better post-surgery and hospitalization period, and distinctly lower blood-pressure readings. The psychologists found that male patients didn't react as well to the touching because, they reason, males are simply not as comfortable with being touched.

ITEM: Dr. James Lynch, director of the psychophysiological clinic at Baltimore's University of Maryland School of Medicine, says, "In our crazy society, we have to *prove* that human contact is good for you." The psychiatry professor did just that with several studies on coronary-care patients and patients in a shock-trauma unit. In a study of sixty-two coronary-care patients, thirty-one of whom exhibited types of cardiac arrhythmia, the frequency of ectopic beats changed significantly when a nurse palpated the pulse. In another study, the heart rate of patients who had been seriously injured and were on respirators showed significant changes when a nurse comforted them by holding their hands. Human contact, conjectures Lynch, acts as a potent stimulus for changes in cardiovascular functions; they can occur even when patients are unconscious or comatose! Further exploration must be made, says Lynch, who hopes these studies will "shed light on why the prolonged or sudden acute absence of human contact seems to have such deleterious effect on heart functioning." In a surprising number of cases, continue Lynch and his associates, "among single people, including widows, unmarried adults, and those individuals who are divorced, the frequency of death from various types of cardiovascular disease rises well above those levels seen in married individuals."

ITEM: Touch heals only when it feels good at both ends of the touch. Dr. T. Berry Brazelton, renowned pediatrician and Chief of the Child Development Unit at Boston's Children's Hospital, cites research about the quality of touch by Dr. Peter Gorski of San Francisco and Dr. Jerry Lucey of Burlington, Vermont. They discovered that rough handling or pinpricks sent oxygen levels of newborns down and gentle strokes and soft touches sent them up.

ITEM: Dr. William E. Whitehead of Johns Hopkins School of Medicine says that just placing a hand on someone's wrist is likely to lower his pulse rate. And, if you touch a person who is in pain, the physical contact minimizes the extent to which the pain drives up the heart rate.

ITEM: Sister Elizabeth Kenny was an Australian nurse. When everyone else was telling polio victims to remain absolutely still, she was massaging, manipulating, touch, touch, touching her patients to their enormous advantage, it was discovered. Sister Kenny also insisted on hot, moist applications as well as passive exercise—two more touch approaches. And no one should forget that Franklin Delano Roosevelt credited much of his own regained body strength, after polio, to his wife Eleanor's insistence on massage.

ITEM: In the journal *Practical Psychology*, I. R. Milberg reports research that shows that the touch of the human hand is beneficial to children affected by skin disorders or diseases. Some dermatologists, says Milberg, even recommend that a mother apply medication with her hand rather than a cotton swab so the child feels a caress instead of a stick.

ITEM: In *Pediatrics,* Dr. Maurice J. Rosenthal reports tests of the theory that eczema can come about in predisposed infants because they fail to receive enough cuddling from their mothers. He investigated twenty-five mothers who had children under two with eczema and determined that, indeed, they were touch-deprived children.

ITEM: Touch is paramount in importance when comforting the very ill, says Stephanie Matthews Simonton, a renowned health and psychology practitioner in the field of cancer. Talking is important in conveying love and support to very ill patients, but, says Simonton, "nonverbal communication is sometimes more effective." Simonton cautions that there may be some people who don't want to be touched and their wishes must be respected. She also reminds us of "patronizing" touch—the perfunctory pat on the back that "infantilizes." Another not-wonderful-at-all touch is the one that stops a patient's outpourings. "I'll hold you but I won't listen" is what it says.

There are, says Simonton, many touch preferences. When comforting a sick person, it's important to find out what feels best. "Some people," she says, "love to be hugged, while others prefer a simple kiss on the cheek, and still others feel most nurtured when their hair is stroked. . . . Two people may like back rubs, but one likes to have his back stroked lightly, while the other likes a good strong rubdown." So, she says, find out from the patient what he finds most comforting through occasional questions, like, "How does this feel?" or "Does this feel good?"

Sometimes, Simonton points out, the support person needs touch from the patient and that can do them both good. If touch is not forthcoming, one must ask for it, and that includes men needing touch from other men, a traditional taboo.

ITEM: Dr. Robert Rosenthal, Assistant Clinical Professor of Orthopedic Surgery at the Harvard Medical School, and Chief of Orthopedics at the Cerebral Palsy Clinic at Boston's Children's Hospital, says that touch can be an aid to youngsters with cerebral palsy. "By overwhelming the sensory input—which is what happens when you brush the skin lightly—spasms are often noticeably decreased. Sensory stimulation is a valuable therapeutic modality, not for all, but for selected patients. Light, gentle touching stimulates the nervous system into trying to function as normally as possible, and it often produces more normal muscle action and improved coordination."

ITEM: As early as 1812, broken bones were treated with electrical impulses, and today it is absolutely recognized that electricity promotes bone healing, although *how* doctors are still not sure. Is not the energy that comes from all living things electric energy? And, if people give off electricity, could it not somehow be channeled from one to another to heal the weaker of the two? That is exactly what therapies like Therapeutic Touch claim.

THE SELF-MASSAGE

On a scale of 1 to 10, self-massage can never be a 10 but it can hit a 7 or 8. The *best* is if a capable someone else gives you a massage, but if there is no available willing soul, why should you have massage deprivation?

Dr. Mike Samuels, the co-author of a wonderful self-help book called *The Well Body Book,* tells us to map a tour of our naked body, to observe it in front of the mirror, and find the beauty in it—even if it's not a Fonda or Schwarzenegger body. Then, he tells us to touch it from top to bottom without losing fingertip contact; hair—feel it with every finger; base of skull, temples—caress each part lightly; then forehead, eyes, mouth, jaw, neck, breasts—massage the breasts lightly; rib cage—all the way down, trying not to miss a place. After you've made the body tour, says Samuels, stretch, lean forward, and touch the floor with your fingertips. "Go easy, be good to yourself." Now, start the actual massage, since you've made friends with your body. Follow the body tour again, use your fingers caressingly, or vigorously, or circularly as you pay attention to each part. Caressing the body will stimulate sense receptors in the skin and increase the flow of energy throughout the body. Vigorous and circular touch will help you relax and open small capillaries under the skin, which in turn increases circulation throughout the body. Watch yourself, suggests Dr. Samuels, as you massage and enjoy the sight of your hands moving over your body. Finally, take the body tour again and use firm pressure with your fingertips, pressing deeply into muscles to relax them. You are simulating Chinese massage now, helping to unblock energy flow. Combinations of techniques, suggests Samuels, are terrific. A self-massage is particularly helpful when you're ill. It opens capillaries and makes it possible for "health-bringing nutrients to get to cells more quickly."

Self-massage during illness also revs up the speed at which waste products excreted by the cells move out of the body, and finally, by opening capillaries, self-massage makes it simpler for antibiotics or other medications to travel through the body and arrive at the infection area.

Exploring yourself with touch does something else. It gives you a familiarity with your body that few people possess. We learn to pick up signals of fatigue and ill health. We learn to prevent illness before it even happens, when our bodies give us early signs that all is not well. We learn what gives us pleasure and what really doesn't without waiting to learn it from a partner, purely by accident.

Gurney's Inn is a spa that sits on the tip of Long Island at the edge of the Atlantic Ocean. Masseurs there recommend a self-massage that sounds wonderful and sloughs off dead skin cells, to boot. In a health-food store, buy some Dead Sea salt and some almond oil and make a paste of the two, says the inn masseur. Massage the mixture over your body, starting at the feet. Always start, say Gurney's experts, with the body part farthest from your heart and rub in a "towards-your-heart" motion. Use a bath mitt, sponge, or even the palms of your hands, and rub up gently. Then, from the neck, rub downward. In a shower, allow the warm water to massage you some more as it rinses off the paste. Finally, switch to cool water to close up opened pores, and, if you like, moisturize with almond oil afterward in another gentle self-massage. Those allergic to iodine should skip the sea salt but, by all means, do the massage.

Self-massage is an act of self-renewal, available any time. Your hands have the power to heal and invigorate, and this is a gift that can be fine-tuned with practice.

Sometimes, in our society with its puritan heritage, we feel uncomfortable ministering to ourselves. Self-attention—especially with touch —can seem too self-indulgent and carry overtones of masturbation—an unfortunate "no-no" from which we are still not free.

In fact, we often have to find situations or places where permission for self-massage is tacitly given—the permission to make ourselves feel good. Such a place is a health spa, and spas exist in small towns and large —feel-good places where massage, either by oneself or another, is acceptable.

Natalie Allon is a sociologist who has studied urban lifestyles in various settings, and one of the urban gathering places that seems to be most popular is the health spa. Why? Allon postulates that it is a place where people can get in close touch with their physical selves and just be "body involved." One young woman puts it this way: "It is as if you are taking care of yourself here like your mother used to take care of you when you were a baby. You mother yourself. . . . It's like hugging myself all over."

And that seems to say it all, for self-massage. It is a temporary way out of a pressure-cooker life. It is a touch that you need no one else to give. It is a mini-vacation.

CHAPTER SIX

FAITH HEALING

First of all, let me state my position. It's this: On the subject of faith healing, I'm fascinated but largely unconvinced.

Next, let me give you David Ben-Gurion's words, expressed in another context but applicable here: "Anyone who doesn't believe in miracles is not a realist."

Last, let me quote from the bard. "There are more things in heaven and earth, Horatio, than are dreamt of in your philosophy."

Now, let's talk about faith healing, which, it appears, doesn't need my confirmation to work.

Faith healing takes its roots from many Biblical descriptions of healing, as in the following from James 5:14–15: "Is any sick among you? let him call for the elders of the church; and let them pray over him, anointing him with oil in the name of the Lord: And the prayer of faith shall save the sick, and the Lord shall raise him up; if he have committed sins, they shall be forgiven him."

Although it's not terribly difficult even for a skeptic to believe that a holy person could touch and heal, the talent doesn't often seem to be gracefully transferable. What has given faith healing a bad name, I think, is the ones who wheedle, exhort, exclaim, and appeal to me over the airwaves, on television, in newspapers, under tents; the ones with the gypsy countenances or the inspired, oh-so-earnest faces, the oh-so-earnest crazies. They do tend to abound, and any of these who tries to lay a faith-healing touch on me or mine, I must say, I'll push. Away. As fast as I can. As much as I love touch, it is awfully hard for me to buy the typical faith healer who tells the sad little lady on national television to throw away her crutches—and she does. Has it been staged? Will it last, this crutchless state? I never know.

And yet. There apparently is little question that faith—that is, what one believes in, whether it be witch doctors, surgeons, psychics, or television exhorters—has been shown to have an enormous effect on the outcome of disease, and I mean a chemical, measurable effect as well as an apparent cure. The placebo effect, once ridiculed as "just a placebo, a palliative for hypochondriacs," is now being credited with tremendous power. If you believe in tetracycline, it may well be the belief that cures you rather than the actual chemicals in the antibiotic. If you believe in a shaman or in a friend or in Lourdes, that also may cure you *no less effectively* than the tetracycline. Even a worrier like me cannot dismiss the mind's power in believing, imaging, deciding— and its real, not imagined, effects on the body.

Lately, I seem to be in the distinct minority when it comes to disbelief that the faith healer's touch will heal. People for whom I have great respect—my best friend, the lawyer down the block, the movie star with intelligence, many of the medical researchers who previously wouldn't touch touch with a ten-foot pole—are suddenly knocking at the door of the faith healer.

Certainly there's a tradition that gives her a place in history, if not science. The faith healer used to counsel kings, touching away their gout, baldness, impotency, and indecision—or so the kings thought. Faith healers were never despots or murderers—that can be said for them. On the contrary, most faith healers advocated nothing more than the sacrifice of a symbolic calf and many didn't even need that.

With all my disbelief, can I entirely negate the work of people like Edgar Cayce, the gentle, shy twentieth-century healer, who never asked for his gift, never charged a penny for it, but went on healing through touch and even through long distance to his and everyone else's stupefaction?

And wouldn't it be arrogant to simply ignore the work of the Brazilian healer Ze Arigo, who died in 1971 at the age of forty-nine? Totally uneducated and almost illiterate, Arigo, like Cayce, never charged for his services, saw hundreds of patients a day, and attributed his power to a voice inside his right ear, a voice he said belonged to a spirit guide named Dr. Adolfo Fritz, a German doctor who died during the First World War. It would be arrogant because, hard as it is to believe, Arigo (through Fritz's spirit voice) treated, wrote medical prescriptions which he had no way of understanding, and prescribed some drugs that were so obscure that most physicians—let alone an uneducated peasant— would never have heard of. But the drugs existed. The voice in Arigo's right ear told him what to write on his prescription pad. The patients, most of them, became well. Arigo operated from time to time. His method was enough to strike fear in the heart of the bravest. Anesthesia or sterility was not in his lexicon. His tools were what were at hand: rusty knives, a pair of garden scissors, a kitchen knife. In seconds or minutes, he removed tumors, diseased body parts, and cataracts, while others observed. The patients didn't complain of pain or anything else. They prospered with health. Arigo was even filmed and those films exist.

And what about the Lourdes cures and their medical assessment put out by that austere and very proper organ, the *Journal of the Royal Society of Medicine*. In August 1984, the *Journal* reported on the work of CMIL (the International Medical Committee of Lourdes), which consists of representatives of major nations who are regularly sent to examine the cases of the Lourdes pilgrims. The committee's mission is to determine if the pilgrims were truly sick, then truly cured. The committee must then determine if the cure is scientifically explicable and if the disease was organically serious as opposed to psychologically debilitating. Finally, the committee must determine whether the disease has actually been cured and is not in a natural remission. Diseases must be said to be absolutely unresponsive to medical treatment or time to make them eligible. If the committee should find such cases, the committee wants it known that the very fact that an apparent cure is "medically inexplicable doesn't make it a miracle because that is a matter for the Church, not doctors."

Not easy to find a miracle. No one ever said it was. However, such medically inexplicable cures indeed were found, the last one documented being the case, in 1982, of a child from a village on the slopes of Mt. Etna in Sicily. Cancer was rampant in her body. She spent four days praying and touching the shrine at the Lourdes grotto. Funeral arrangements were completed. She never made that funeral. According to the committee, prayers and touch seem to have wrought an inexplicable cure. No one over the age of five had ever before experienced a spontaneous remission of her particular disease.

Other cures have also been accepted as real but medically inexplicable in the thirty years since CMIL was established, reports the *Journal of the Royal Society of Medicine*, almost reluctantly. Despite much of the medical establishment's reluctance, something like faith healing appears to be happening.

There is the shaman Essie Parrish. Shamanism is the faith of primitive hunters and gathers through the world. A shaman is a mystical healer who owes her or his powers to the spirit world. The shaman communicates with a spirit, who selects him and whom he cannot refuse. He does this by going into autohypnotic trances. Shamans hold great power and prestige in their tribes, and Essie, the Pomo Indian shaman in California, was so revered. A documentary called *Sucking Power* was made about Essie because that's what she did—she doctored with her mouth and throat, sucking diseases from her patients. Sometimes she used her hands: "When I work with the hand power, it is just like when you cast for fish and the fish tug on your bait—it feels like it would with the fish pulling on your line—that's what it is like."

Essie is very open about her methods and has been widely studied by anthropologists; her system of belief and her own faith-healing work have been noted and validated time after time.

As a skeptical, big-city dweller I say hogwash. The poet in me says, "Dummy, 'there are more things in heaven and earth than are dreamed of in your philosophy.' "

Actually, many faith healers are beginning to infiltrate the mainstream in visible ways. The Reverend Camille Littleton, for example, who leads the healing service at St. James Episcopal Church in Marietta, Georgia, conducts a healing service, a laying-on of hands straight from the *Book of Common Prayer* (page 456)—a service that had been dropped from the book until restored in 1977. At main-line services all over the country, like those at St. Philip's Cathedral in Atlanta, St. Luke's Parish in Ohio, Trinity in Manhattan's Wall Street district, Heavenly Rest on the Upper East Side of Manhattan, and countless other places of worship, healing "fire" has caught on.

It has almost become unscientific to ignore the scientifically unproved effects of the laying-on of hands, among other "unscientific" treatments.

Now that the establishment religions have begun to climb on the faith-healing bandwagon, as writer T. George Harris in *American Health* says, a "productive tension should result. It could help take medicine beyond its sterile, dehumanized theory, and bring faith healing out of the Middle Ages."

Even doctors have begun to initiate hands-on prayer if the patient seems receptive. They do this at many hospitals and clinics like the Princeton Medical Center, where cardiologist William Haynes, graduate of Princeton (1950) and Columbia Presbyterian Medical Center (1953), prays with his patients in the coronary care unit.

In fact, more and more of the medical establishment (but by no means all) has become interested in mind-body connections, and there are traditional physicians of impeccable credentials who unequivocally state that there is no disease that is *not* psychosomatic. *Psychosomatic* is a word that has received a bad press. Its meaning has been muddied

by wrong usage. All *psychosomatic* should connote is "mind-body"—
that both body and mind are affected, nothing more and nothing less.
It does not mean that the illness is pretended, not serious, not real in
the very physical sense. Nor does it mean *psychogenic*—the word for
an illness that is thought to be *caused* by a disturbance in the mind or
psyche. It only means that the nonphysical and the physical, the mind
and the body, are both involved, interacting with each other and ac-
centing each other. We are not simply bodies—we are mind-bodies,
says Dr. Andrew Weil, research associate in ethnopharmacology at Harvard University in Boston, Massachusetts.

Doctors everywhere are beginning to recognize that the mind
affects a disease—and its cure—in profound ways never before thought
possible. More and more are beginning to look harder at "the positive
placebo effect." This is what happens when a patient takes a "medically
inert" (without useful chemical properties) substance or undergoes a
medical procedure with no known intrinsic therapeutic value and,
nevertheless, the substance or procedure works. The patient gets better. Clearly, the patient's renewed health had to do with the connection
between her mind and her body, and if faith, belief, or will works, why
put it down?

Dr. Herbert Benson, Associate Professor of Medicine at Harvard
Medical School and author of *The Mind/Body Effect,* says, "in its present disregard for the *positive* placebo effect, medicine has lost a valuable asset, an asset which sustained it for centuries."

Hippocrates made similar remarks a few centuries before Christ.
"Some patients, though conscious that their condition is perilous, recover their health simply through their contentment with the goodness
of the physician." And there is a lot to be said for the healing properties
of a good doctor-patient relationship. It surely is one form of the "placebo effect."

Dr. Bernie Siegel, who practices surgery in New Haven and
teaches at Yale, says that "one of the best ways to make something
happen is to predict it." The placebo effect, he says, has been ridiculed
for twenty years by traditionalists, even though the fact remains that
about a fourth to a third of all patients show improvement if they
merely *believe* they are taking powerful medication even if the medication they take is virtually worthless. Today, says Dr. Siegel, most of the
profession has stopped ridiculing.

Faith healing, which is a mixture of trust, hope, and belief—both
the patient's and the doctor's (it helps if the healer, as well as the
patient, really believes in his power)—for the time being, at least, tran-

scends intellectual reasoning (which is why my own skepticism frequently wavers when I see another example of success). More often
than not touch seems to be behind many of the faith-healing successes.

Who is so brilliant and all-knowing that he or she can say for a fact
that there is no magic-like touch? For one thing, we are learning new
things we *can* add as pretty convincing evidence to our repertoire,
almost daily. Take ESP. We never knew until recently that ESP (extrasensory perception) actually existed. But there is such a thing that many
call a "sixth sense," and documented studies at Duke and other prestigious universities have shown that there are those who, blindfolded, can
name objects or colors that are being pointed to in another room. No
tricks.

And there are people like Maya Sanders, who dreamed that her
relative was in an automobile accident and, the moment she was dream-
ing it, he was. One might chalk this up to coincidence if it weren't for
the fact that this sort of thing happens all the time to Maya. There are
people who get clairvoyant vibrations, and if they get them, why can't
they give them in the form of healing? Why? Because the head tells us
they can't. But the heart, at least my heart, continues to hope my
mother's stroke can be healed if she only believes in the little woman
who says she's sure she can, with her talented hand, stroke away the
worst part of the stroke.

There is something in me that (I think) is wise, that says: Don't
throw the baby out with the bath water. People who are so prejudiced
they say every extraordinary thing that occurs *must* be a hoax or a
mistake or a coincidence are as non-seeing as people who are so gullible
they believe *every* extraordinary thing that seems to occur.

Dr. Jerome D. Frank, Professor of Psychiatry at Johns Hopkins
University, analyzed the components of the healing procedures in
primitive societies, and his studies concluded that the healing effect of
these procedures derives from the patients' expectations of help. Expectation of help can be potent. It has been referred to as "expectant
trust," and it results partially from the healer's personal characteristics
or charisma.

If healing methods reside in the patient's state of mind, and a faith
healer can work up such a state of mind, and the healing is effective,
why is it less good than penicillin? I am forced to ask myself.

Charcot, the nineteenth-century French neurologist, once said that
"the best inspirer of hope is the best physician."

"Medicine in the Western World," says Dr. Benson, author of *The
Mind/Body Effect,* "which was once an integration of both science and

art, has shifted to an emphasis on science at the expense of art." Perhaps the faith healer's work, which can be defined as a kind of art which employs a bit of belief, a little magic, a little street smarts, and a little sixth sense should not be so swiftly made an object of derision. Perhaps there is a place for faith healing in an age of questioning and disbelief, an age when religion is fast becoming as outré as magic. Perhaps where people, in hope, once turned to rabbis, priests, and holy books for answers about how to become virtuous enough to deserve health, there is a place today for psychic healers who may still give hope for health to the religiously disenfranchised.

Today, even science seems to have little question that the mind and the body intimately interact. As I've noted, the whole Cartesian notion of mind-body dualism is being discarded. Moreover, there is also little question that many serious people all over the world believe that psychic healing is possible. In 1975, at London University, the World Federation of Healing was established, and one of its purposes was to make faith healers available to patients everywhere, as they are in Great Britain. There, the Minister of Health has granted permission to the British National Federation of Spiritual Healers for its members to see patients who ask for them at about fifteen hundred hospitals throughout the country. The federation, which claims about two thousand active practitioners, is certain its methods work but few make an attempt to explain how.

In the well-established Tenrskyo Hospital in Japan, there are about eighty healers available to serve patients for whom "all hope has been abandoned," and Dr. Toshio Yamamoto, the director of the hospital, has claimed that many of these patients have been cured by the faith healers.

Some medical researchers say that some 80 percent of all human ailments are psychosomatic; others say 100 percent: We *are* mind/bodies. Belief or faith, then, is very powerful. Could it hurt to reeducate our minds—not to exclude the miracles of technological medicine but to include the miracles wrought through the art of faith or psychic healing? A touch by such a healer may touch something in the human spirit beyond the realm of science, beyond even the present imagination of science.

The psychologist Lawrence LeShan, author of *The Medium, the Mystic and the Physicist,* says, "Coincidence has a long arm and the unexpected does often happen. All reports of medical improvement by psychic healing (or by any other therapeutic technique) must be interpreted with this in mind. It is also true, however, that a technique must be evaluated in terms of its results, and not in terms of a previously held theory."

That last sentence says it all. If you are ill, and a healer touches you and you get better, wouldn't the open and educated mind rethink her thoughts about psychic healing? It *could* be coincidence, but maybe, just maybe . . .

Through all arguments, pro and con, the fact remains that we do not yet understand how some things work. Doctors do not yet understand just how aspirin works. However, that is a poor reason to throw out the aspirin bottle or to deny the existence of other unexplained healing techniques—perhaps as short-sighted as it would be for a New Guinea tribesperson to deny the existence of television, Polaroid cameras, microwave ovens, Saranwrap, or David's Cookies.

Dr. Sidney Burwell addressed an entering class at Harvard Medical School saying: "Gentlemen . . . medical science is progressing so rapidly that by the time you will have finished your four-year course, one-half of what we tell you will have been by that time proven incorrect and unfortunately we cannot tell you which half it's going to be."

Perhaps next semester someone will figure out how and why faith healing works and how to reconcile faith and modern medicine.

I said, as I began this chapter, that I was largely unconvinced. And yet I seem to have made quite a case for faith healing, haven't I? Hmmmm. Perhaps what I have inherited—a stubborn belief in the invincible wisdom of the chief of service of the finest technological hospital—is nothing more and nothing less than faith medicine. Perhaps a greater willingness to take control of my own health might lead me to a fuller acceptance of the part my mind plays in the well-being of my body. I'm willing to check it out.

CHAPTER SEVEN

TOUCH TACTICS

Truth is as impossible to be soiled by any outward touch as the sunbeam.

JOHN MILTON

THE POWER PLAY
(Touch as Silent Force)

She was twenty-six and just reporting for work as a newly hired attorney in a large Washington firm. The morning went well, she was introduced to all, and later, standing in the hot-foods line in the company cafeteria, a young male associate, three years her senior, took her elbow and steered her over to the salad bar. "This is much better than the other garbage," he announced, still holding her arm. "Can you fix her a plate of the freshest leaves?" he asked the attendant.

Her stomach felt tight, his touch burned her elbow, she resented it enormously but couldn't pull away. That would be rude.

"Stop being a crazy," she silently told herself. "He's just trying to be nice. He doesn't know you hate salad." Still, the gentle pressure rankled. His fingers felt like an invasion.

The senior partner appeared. He put his arm around the male associate, told him he was needed in the conference room, smiled briefly at the young woman, and then the two men left together, the older with his arm still draped over the younger man's shoulder. Touché. Touch at work. Rank ranks.

Power speaks eloquently through touch tactics. The messages are varied. Among them:

· I'm in charge here.
· Women count less.
· Age pulls rank.
· Youth is stronger.
· She's my sexual property.

143

Who would think of putting an arm around the doctor? The doctor's patient? No. His nurse? No. His secretary? No. The lab technician? Double no. Could the doctor, *would* the doctor easily put his arm around any of these people in his life? Yes, yes, and yes.

In doing so, he would be invoking a system of touch privilege that would be understood if not totally appreciated by all. Touch, says Dr. Nancy Henly, who has made exhaustive studies on the politics of touch, can be "as friendly as peanut butter" and yet an absolutely clear sign of someone's subordination to another. Superior-status people, Dr. Henly reports, are much more likely to touch inferior-status people than the other way around. There is a hierarchy of touch in which one can always touch the person a step below on the ladder, while it would be unthinkable for the occupant of the lower step to reach up and touch back. Insidious and subtle tactics are at work, and what appears a friendly, democratic gesture may actually be a strong statement: "Don't forget who's the boss." Whole social pecking orders are contained or maintained by the power play of touch. Nobody would mistake the touch of the young male associate in the law firm. Nobody could fault the young woman for resenting his touch, which implied at least three things very loudly: I'm senior here, the touch said. I'm a man and therefore stronger and taking over here. Also, I know what's best for you. And finally, as a man, I have dibs on you. And surely nobody at all could mistake the power touch of the senior partner, letting everyone know who was really boss.

Dr. Henly's studies indicated that dominant and subordinate positions in life changed depending upon the situation, and touch was the giveaway as to who was in the dominant and who the subordinate position. Someone, for instance, who gave "information, advice or an order" would be in a dominant position and more likely to touch than be touched. Somebody who was trying to persuade another to do something might well try to assume a dominant position by touching the object of his entreaty.

Consider your dating days . . . or those of your children. Despite the rise of feminism in many modern cultures, it was and still is the male who is generally the first to touch his female friend in a romantic way. Who in the couple is the first to sling an arm over the other's shoulder, hold the other's hand, stroke a face? The male. Who is traditionally the first to initiate more serious sexual activity? The male, that's who. Who is, almost always, the first to make the first phone call after the first meeting in a kind of symbolic touching facilitated by the long reach of a telephone wire? The answer is, almost universally, the male, whose maleness and power are explicit in a thousand different kinds of touch-ings.

The handshake is the caveman's legacy to subsequent generations. It is an example of touch, not only as contact, but touch as contract—a touch that says, We're equal here and we can trust each other. Dr. Henly says that the handshake is essentially a masculine ritual that implies male "clubbiness" and one that tends to "exclude women from the club." Little girls, tired of being excluded, seem to have picked up on the touch-as-contract notion by employing the pinky-link, a femi-nized version of the handshake, in which two little girls seal a bargain by intertwisting their littlest fingers. Very few males even under the age of nine have ever been caught in the midst of a pinky-link, which is sillier, cutesier, and somehow less binding than a real handshake. When women of today who have outgrown the pinky-link meet for the first time, they, like men, do tend to shake hands. As they become more friendly, though, unlike men, they begin to eschew the handshake in favor of a limply placed kiss, usually somewhere just to the left of each other's ear. It is, like the pinky-link, a form of touch but a less dynamic form. Finally, when a woman meets a business or social friend who is male, none other than Emily Post suggests that she retain a more pas-sive role than he and do actually less of the handshaking that occurs in favor of just standing there graciously, allowing her hand to be pumped. Lip service to equality is often belied by the real power label—the exchanged touch. Conversely, when males meet for whatever purpose, each is usually as vigorous as the other in a handshake. *If* one male doesn't pull rank.

Make no mistake: the handshake is a powerful tool of domination. Allan Pease is a communications expert in Sydney, Australia, and he's made a study of dominant and submissive handshakes as power plays. Turning your hand so that your palm faces down as you meet another hand transmits the idea that you intend to take control in the encounter that will follow. Pease reveals that in studies of 54 successful senior-management people, 42 actually initiated the handshake in this palm-down, dominant position. On the contrary, if it's to your advantage to let the other person know you're giving up control, you'll offer your hand in a palm-up position.

Have you ever been a dominant shaking hand with another domi-nant? "As each person tries to turn the other's palm in a submissive position," notes Pease, "the result is a vice-like handshake with both persons remaining in the vertical position as each person transmits a feeling of respect and rapport to the other." This is the "shake like a man" handshake fathers generally try to inflict on their son's education.

It gets more complicated, the handshake touch does. There are ways to counter the dominant handshaker, the palm-down thruster, and the "shake like a man" advocate. They involve a complicated choreography of fancy stepping and disarming with the other hand or using the "glove shake," a ploy often known as the politician's handshake, whereby the politician encircles the shaker's one hand with his two. Pease discusses "knucklegrinders" (tough guys), "stiff-armers" (those who want to keep you at a distance), "wrist-holds" and "upper arm grabs" (more power plays) as well as "body lowerers" (those who bow as if shaking hands of royalty).

The self handshake is a powerful message. That's why every public winner in the world raises his hands high above his head and clasps them. Sometimes, if he wishes even more visibility, he waves his clasped hands.

The black militant also gives an unforgettable power message as he raises his hand, fingers bent to touch his own palm in the classic fist.

The "little push" is a wonderfully powerful touch ploy. President Lyndon Baines Johnson always looped his arm around the shoulder of a visitor in what the visitor was supposed to read as friendship. No matter what the visitor read, the real message was "It's time to go." In this manner LBJ always little-pushed people out of his presence when he was finished with them. Many chief executive officers of businesses do the same. Leonard Lauder, chief executive officer of Estée Lauder Inc. looped his arm around my shoulders in a gesture of the nicest, friendliest sort just the other day; before I knew it, I was on my way—and not even feeling angry, so tactful was his "little push."

Touch as a power play can often be seen in a New York subway. There, it is a kind of sign language for insistent force. Some males will openly run a lingering hand along the thigh of a woman who is trapped in a crowded car as she submits angrily but essentially helplessly: before she can yell or complain, the act is completed. The man with the roving hand realistically knows he cannot endear himself to the woman by doing this. His hopes for future sexual contact cannot be strong—if he is sane. He cannot derive great sexual pleasure from the errant and disconnected touch. All he can get is a sense of power. He has humiliated by his touch and that makes him stronger.

To many, touch implies dependency and the toucher asserts his power by insisting on that dependency. Infants and children receive loving, but still power touches as they are diapered, dressed, fed, bathed by adults who are bigger and stronger. One friend remembers the

despised touch of her father, "the neck hold," she called it. He'd encir-
cle the back of her neck with a viselike grip as they crossed the street,
sure that he was protecting her in the most fatherly way; she only
remembers the implicit message of that neckhold—"I'm boss," it said.
Adolescents often reject parental touches because they are rejecting
the power of their parents to influence them. And too many adults, in
what should be the prime and the power of their grownupness, still shy
from touch because it reminds them of being dependent. How many
males refuse to be cuddled and certainly refuse to cuddle, except in a
purely sexual situation? Their partners, puzzled, attribute it to coldness
or lack of interest, when perhaps the overt touch is seen as a power play,
not as affection.

And when one is normally an affectionate person, and suddenly one
withholds touch? Talk about power plays. Strong men have been
brought to their knees by women who submitted coldly to the sexual
acts required by marriage contract and yet refused to give the casual,
loving, nurturing touches to which their partners had become accus-
tomed. "I'll do anything—just touch me again" is the unspoken plea
such a power play evokes.

The same touch feels different to different people. A man patting
a small child on the backside implies loving acceptance. A boss patting
a secretary on the backside implies . . . oh, terrible things. A husband
patting his wife on the backside can imply warmth or dominant power,
depending on their relationship. A coach patting a player on the backside is imparting his powerful good-luck charm. Ashley Montagu has
said that cheek and hair patting and chucking under the chin are all
"forms of behavior indicating affection." But not always, Dr. Montagu.
The bully down the block can pat the hair or cheeks of his victim and
give a message very different from affection. The bully's chin chuck can
be less than affectionate, even though it may not be any more violent
or physically hurtful than a daddy's chin chuck.

Touch establishes territorial lines and nowhere are those lines more
clearly drawn than in sex. Watch a young man talking to a newly met
and desirable young woman at a cocktail party. If both are sitting on a
couch, for instance, his body will be turned in toward her and he will
no doubt be fingering her knee, her shoulder, her purse—something on
her or something that belongs to her in order to give the message "Stay
away—she's taken."

Not long ago, Elizabeth divorced her husband. An architect, she
worked in an office where most of her colleagues were male and married. Before her divorce, she and her husband spent many hours social-

izing with those colleagues at jointly given dinner parties. Always, there had been post-dessert discussion between the architects, who loved their work; the various spouses put up with the business chatter with good-humored complaints.

The week after Elizabeth's divorce became final, she was invited to yet another dinner party, and after the dessert she and a colleague settled down with their coffee in the living room to chat as they'd done on myriads of other evenings. But not, apparently, on this evening.

"Suddenly, out of the blue," remembers Elizabeth, "his wife, also a friend, bore down on us and *literally sat on his lap* in such a way as to totally block our eye contact. His neck *bent* with the sudden weight of her arm as she flung it around him. If it hadn't been so sad, it would have been funny. She touched him with every part of her body in such a proprietary manner that the message was unmistakable. My divorce made me a threat, and the conversation, as well as the friendship, it turned out, were finished."

There was something else going on here. Elizabeth's friend's wife was, in no uncertain terms, establishing what has been called the prerogative of "intimate distance." Distance establishes power structures, and that is why, in most societies, people very carefully use and apportion to others very specific amounts of space. Touch, more than first names, more than pedigrees, reduces or enlarges social distance.

Over a decade ago, Dr. Edward Hall, a student of human behavior, made a study of people's use of space. He determined that there were four categories of interpersonal space and they were:

- *Intimate* (from direct contact to between 6 and 18 inches, the amount of space the architect's wife put between herself and her husband to send a message to Elizabeth)
- *Personal* (1 1/2 to 2 1/2 feet: good friends)
- *Social* (4 to 12 feet—cocktail-party distance?)
- *Public* (12 to 15 feet; which is why crowded subways feel so oppressive and threatening)

Dr. Hall made the point that anyone who imposes intimate touch contact in a social or public place is breaking tacit rules. People protect their space from touch intrusion in various ways. At cocktail parties, they hold drinks in front of them, silently warning anyone who would touch-trespass, to beware. At meetings, participants often stand with arms crossed and significantly jutting out in front, warning: don't touch. A good friend who puts her face just a whit closer than one and a half feet from yours, as she relates a story, makes you somehow uncomfortable. It's just not cricket, just not fair to overstep boundaries and take unfair power advantage in doing so. That is the unwritten law.

Besides space invasion, power touch can be asserted by a "pompous, superior-to-thou" gesture, says sociologist Natalie Allon, who made a comprehensive study of singles bars across the country. If one waited long enough, said Dr. Allon, in almost every bar would eventually appear, each night, the Specimen. He usually looked like a clothed *Playgirl* centerfold, and he invariably walked across the room literally pushing people away as if he were saying, "Watch out, I'm coming through and I'm hot stuff." This man, says Dr. Allon, wanted to be sure that every woman noticed him. He wanted recognition for his virility and body potency, and the surest way he could get it was to manipulate physically those in his path—not enough to get a sock in the chin but enough to make a point.

Sometimes one applies to a perfect stranger for a service that only the stranger can provide. In order to provide the service, the stranger must touch you, actually physically manipulate you. Even though you've asked for it—as when you visit a doctor—the power suggested by such touch intimacy, even "allowed touch," can be disturbing if the touch isn't given in prescribed ways.

Dr. Richard Heslin of Purdue University agrees. The power of touch can be off-putting even when you need it, he comments, and this is particularly true in relationships between professional people and their clients; it's virtually a "manipulator to object" relationship, says Heslin. The power always falls to the manipulator. If the touches from a doctor to patient, golf professional to student, tailor to customer are disturbingly intimate, the receiver of the touch is at a distinct power disadvantage. That is why, says Heslin, such professionals often seem rather businesslike—even cold—so that it's clear that the relationship is not personal but professional-functional. In our society, we often decry the too-cold manner of certain professionals, but the coolness is often necessary, notes Heslin. Moreover, the license for such professionals to touch is strictly limited. The shoe salesman can play with your instep but he'd better not let his hands wander further up your leg. The hairdresser can fondle your scalp in the most intimate way, but he'd better not touch below the shoulders. Dentists, for the most part, must allow their fingers no more access than to the oral cavity, although an empathic touch on the shoulder after a particularly painful session might be acceptable—even welcomed. The underlying motive and the use of good judgment in each situation are all important.

Moreover, as our Victorian squeamishness begins to recede—along with the implication that all touch between adults somehow implies sexuality or control or both—Heslin's analysis may require change. More and more doctors, nurses, and others in the helping professions are coming to understand the vital importance of real, empathic touch and are tramping across old boundaries. Psychoanalysts, like Rufus Peebles of Cambridge, Massachusetts, are bringing touch into therapy sessions with patients (see next chapter). On a more conservative front, more and more use touch for comforting and healing—not just for professional investigation or treatment. (See *Medical Massage*, page 126.) Through the years, nurses, perhaps, have been freer to use touch than most professionals; their duties required that they touch many parts of their patients. With such close contact, they could sense how and when to use comforting touch as well. And today nurses are frequently given special instruction on how to use touching in their daily rounds—quite apart from such specialized protocols as Therapeutic Touch. With the rise of technology and the specialized machinery they must deal with, nurses, too, have lost some of their ancient art and are seeking to regain it.

However, there is no question that touch, because it is so powerful, can be interpreted in many different ways.

The toucher needs to take great care about whom he touches, how and in what context. What is given as a friendly touch can easily be construed as a put-down. Lynne Leavy, a New York psychoanalyst who works with dwarfs, among other people, notes that many of the "little people" in the prime of their adulthood are often swept up and actually lifted off the floor in displays of affection, much as a child might be. "An adolescent who is three feet tall does not take it kindly when a six-foot-tall person lifts her off the floor—even though affection and exuberance have prompted the act," says Leavy. People in wheelchairs or using walkers are often subjected to friendly pats on the shoulder or head, pats that are taken to be condescending signs of pity or superiority by those so touched. One would never touch a stranger standing in line in a supermarket if the stranger questioned one on price, for example; but, if the stranger is wheelchair bound, a touch might well be given along with the answer. Touch has multilevel impact and it is often construed as a power play even when it is meant benignly.

Sometimes, the opposite is true. A benign-appearing touch can be a power play. The wife giving her husband the sarcastic and patronizing pat on the shoulder that says, "Wonderful, you beat your nine-year-old son in Monopoly—aren't you a hero!" is asserting moral superiority over her doof of a husband. The undertable pinch or the finger-prod touch tells the child, the mate, or even the parent, "Be careful what you say," but because it is performed with a smile to put off the observers, one might well see the finger prod as a touch of intimacy. The finger prod often graduates into the fingernail dig, also a benign-looking, if not benign-feeling, touch.

Those who have ever visited a home for the aged or mentally disabled might have seen certain attendants stroking a patient's hand in a touch ploy that is more dismissive than loving, more restrictive than bars. Such a touch strikes coldness into a patient's heart and confuses as well, because the toucher is often smiling as she disables with a pat. Studies show that even well-meaning nurses can touch badly. Those, for instance, who place an arm around a patient's shoulders cause more discomfort than those who touch a patient's hands or arms, the former being a "take charge" rather than a loving gesture. The touch power plays that look affectionate can be, then, at the worst, malevolent and at the best condescending.

There are certain self-touches that are also unmistakable power plays. High on the list is self-grooming.

When a monkey grooms himself, he's indulging in comforting or cleaning himself, no matter if he grooms alone or in the presence of others. When a person self-grooms herself in the presence of a speaker who is trying to convince her of something, she is wielding the power of dismissal. Gentle self-grooming, picking at one's nails, scratching one's head, picking lint off one's own suit implies that what the speaker is saying is less than compelling. But the more personal and intense the self-grooming becomes, the more of a direct insult and a direct denial it becomes. If a speaker who is face to face with you, or even on a podium, sees you using a toothpick, whipping out a comb to your hair, applying lipstick or eyedrops, or calumny of calumny, picking an ear with an instrument like a matchstick, the speaker knows he's dead. Discreet peeks at a watch are forgivable, but touching the watch as one checks it is a powerful ploy that suggests boredom or impatience.

Folding one's arms as one listens is as off-putting as an arm stretched out to keep the other at arm's length. Silent finger drumming? It kills. Nothing less.

The power plays between the sexes are punctuated graphically by touch plays. What young girl at a beach with her "crowd" has not felt public humiliation if she has experienced being swept up by the males and thrown forcibly into the water? In the fifties, women were even primed to squeal with pleasure as that nonfun act occurred. Waitresses, says the social researcher Nancy Henly, are touched all the time by male customers who would not dream of touching the waiters.

"Women," says Henly, "do not interpret a man's touch as necessarily implying sexual intent, but men interpret a woman's touch that way." Dr. Henly tells of a woman who sat with her husband and a male friend at a party, the male friend's arm being slung casually around her shoulder in what she took to be a friendly gesture. The woman plonked her arm around the male friend's shoulder in the same spirit. Bad news. The instant they were alone, he suggested bed. When she looked at him stupefied, he said, "Wasn't that what you were trying to tell me all evening?"

She had felt his arm was a gesture of friendship or, at the most, a subtle message of superiority. When, as a female, she returned the gesture, friendliness or superiority were out of the question: it had to be sex she was after.

Women have a responsibility to themselves, says Henly, to "refuse to accept tactual assertion of authority." They must gently but obviously remove the hands of those whose touch is not welcome. Men should guard against using touch to assert authority over women and should train their sons to do the same thing. And, if women note that the men in their lives can't or won't break the power-touch habit, they should start touching back.

For good or for bad, then, touch has infinite potential as a power play. It can also be a tool of incomparable persuasion.

The marvelous comedian Mel Brooks has a routine about Torquemada, the legendary cruel Inquisitor-General of the Spanish Inquisition.

"Forget it," says Brooks knowingly. "You could never Torquemada anything."

Well, touch can talk 'em *into* anything.

THE CONVINCER
(Touch Talks You Into It)

Picture the scene:

A public telephone booth. A quarter lies on the ledge just under the telephone. Man enters. Makes call. Pockets quarter. Leaves.

"Pardon me, sir, did you find a quarter I think I just left in here?"

asks the young woman standing near the booth.

The scene is repeated many times.

The answer is always no.

Now picture this scene:

A public telephone booth. A quarter lies on the ledge just under the telephone. Man enters. Makes call. Pockets quarter. Leaves.

"Pardon me, sir, did you find a quarter I think I just left in here?" asks the young woman, as she gently touches the man on his arm for just a second or two.

The scene is repeated many times. When she touches, she gets the quarter back. Almost every time.

Dr. Heslin spends a good part of his life thinking of experiments to prove the potency of touch as a convincer. Here's another one.

Dr. Heslin asks some library clerks to gently touch some book borrowers' hands and then carefully make a point of *not* touching the hands of other book borrowers as they present their cards. Outside the library, he stops the borrowers to question them. Result? Those who were touched for even a split second were far more likely to report positive feelings about themselves, the library, and the clerks who had helped them—even when they were not aware of being touched—than those who had not been touched at all. When you are touched in a friendly way, you seem to sense that you are valued by the toucher. Being valued leads to self-confidence, leads to feeling good. And a person who feels good is easier to convince of something than a person who feels cranky. In short, you have in your own hands the power to make someone feel good about you and himself.

What can this mean? We spend lifetimes developing our verbal convincers. Author Dale Carnegie made a hefty fortune writing *How to Win Friends and Influence People*. And all it takes is touching?

Sounds simplistic, maybe it is a *little* simplistic, but there is hardly a better convincer than touch.

Do you remember sitting in the library with the person of your dreams, studying together? The air is a little tight because you want to go walking after the library and she wants to go to the movies, and the unresolved tension is thick enough to hang heavy over the books. Then she reaches out under the table and touches your foot with hers. It's nothing much, only a foot touch. *Only* a foot touch? The air magically clears. You both go back to the grind with secret smiles and ankles entwined. You go to the movies. You think you always wanted to see that film. You'll walk another time.

Heslin and colleagues Whittier and Abella arranged a study to see just how much touch could convince and there was no arena better than the selling arena. Since the largest element in almost every marketing budget for almost every industrial firm is personal selling rather than advertising, it makes sense for merchandisers to find salespeople who have "the right stuff," an ability to influence customers to buy. When the researchers began, there was only one existing study about the effect of touch on selling. It traced random shoppers in a supermarket who were touched by a saleswoman who asked them to sample a new brand of pizza—and the touching, in that study, was effective; more of the touched shoppers would sample *and* buy the pizza. The Heslin study used three situations. A customer would not be touched at all in an exchange with a salesperson, or a customer would be touched just briefly with a perfunctory handshake, or finally, a customer would be touched several times in a friendly, warm exchange. The findings were fascinating. Customers who were touched (compared to those who were not touched) felt better, liked the salesperson more, felt the salesperson liked them, and felt clearly more influenced by the salesperson. A touching salesperson has the right stuff.

This would come as no surprise to the makeup tycoon of the world, Ms. Estée Lauder.

"I touched, I touched, I never stopped touching," she told me. "When I was building a business, I reached out and touched strangers in an elevator as I 'suggested' they try a little of my cream blush. My father was from the old school. 'Estée—all that touching,' he used to say. He hated it. Thought it was for peasants. But, what I liked to do better than anything else in the world was to touch faces. I have never stopped."

Clearly, Ms. Lauder's makeup has something going for it in purity of ingredients and color, but everyone in the business of cosmetics is convinced that the lady herself is the most powerful ingredient. She was and still is a superb salesperson. She doesn't stop touching.

Touch talks you into opening up for other people, being receptive. Many of us who are teachers know that instinctively. In all the years, in all the classes I've taught, from students who are cerebral-palsied, ghetto-nurtured, or upper-class-privileged, I never met a kid who didn't respond to a touch. The surliest, angriest teenager will pause for a moment, thrown off balance with a loving touch, and let you at least plead your case. Sidney Simon, another teacher-author, recently wrote in the *Reader's Digest* that those youngsters who were rarely touched at home seemed withdrawn, prone to living in a fantasy world, even hostile. "You can see them on any playground—kids who seem absolutely devoted to tripping each other, shoving, wrestling and fighting. Inside the schools, they push each other down the stairs, shove faces into the drinking fountain, throw food in the lunchroom. And behind every shove and trip, there is an unheard cry of skin hunger."

I remember Pete. He was such a seventeen-year-old. He caused chaos in the classroom with his snide asides. "Get the teacher" was his chosen way of life. One day I asked him to stay for a minute after school. He horsed around at the request but complied. On the pretext of going over a paper, I walked to his desk when the others had gone. He looked away, said an insufferably insolent thing—and I touched his face, the lightest of touches, for the briefest of seconds.

I had his attention. He quieted, like a nervous colt. He was approachable—barely. It took weeks of touches to turn him into a doll (well, *almost* a doll), but it worked. I had my class back again.

I learned that touch could bear great influence in the classroom from one of my own grammar-school teachers, whom I've never forgotten. She spent her school day doing her best not to touch anyone, and she avoided being touched as if fingers brought plague. She even stood back when you spoke to her because she "didn't like being breathed on." She wiped off her chair before she sat, she carried her own cake of what she called "lifebooey" to the girls' bathroom, and she remained germless, inviolate, untouched throughout the school year. She was also a lousy teacher. No one ever *wanted* to touch her—let alone listen to her.

I remember teaching cerebral-palsied children and finding that a hands-on approach to those whom strangers rarely touched worked wonders. A little boy who thought he couldn't take a step with his new braces could be persuaded to try if someone would embrace him, hold him, touch him into it. In 1983, a study at Jacksonville University showed that touch was indeed an effective strategy for counselors of handicapped students. The counselors of a test group of such handicapped people discovered that they could effectively reward positive behavior and minimize negative behavior through touch—uncondescending pats or holds.

"For lovers," wrote John Cheever, "touch is metamorphosis." For waitresses, touch is tips—change in its quite literal sense. Researchers April H. Crusco and Christopher G. Wetzel reported an experiment of touch as influence in *Personality and Social Psychology Journal.* The two staked out two restaurants near a small college in Oxford, Mississippi, and asked the waitresses to do one of three things—touch a diner on the palm, touch him or her on the shoulder, or not touch at all, as they returned the diner's change from their bills. Again, the results were extraordinary. Diners who had been touched tipped better than those who had not. An interesting follow-up to this experiment: the physical contact did not elicit any better thoughts of the quality of the food, restaurant, or waitress. It simply elicited greater tips. Some diners, when queried afterward, were not even aware of having been touched. This happens often in touching experiments, which leads one to believe that brief, unobtrusive touching may have great subliminal effects. A touch, in other words, doesn't have to be a hug or a stroke—merely a passing contact.

"Putting the touch" on someone is eloquent even in smallest measures.

THE COMMUNICATOR
(Touch Gives Your Message)

Western Australia is the largest Australian state but that isn't its most extraordinary claim to fame. This is:

Every year to a tiny cove called Shark Bay come, not sharks, but dolphins by the dozens. Children come also, and hordes of grownups and even tourists from across the world, to see these great, galumphing creatures, these aliens from a watery world, communicate with people using the simplest gesture of all—touch. They are natural and wild, not the trained Marineland sort, and they nose around to be stroked and tickled and scratched under their chins. They swim to the shallowest water they can handle where the children can play with them, and they give rides and wonderment in an extraordinary act of communication between very different species. "We're here, we like you, we hope you like us," is the message, and it's heard sharp and clear in the gentle nosings and bumpings that are passed from dolphin to human.

When people touch people, the communication is just as powerful. Freud knew that—but he lived in the Victorian age. "Don't touch" was Freud's advice to his students. It evokes too many sexual feelings, which are then open to misinterpretation. It muddies the analytic waters.

But Dr. Rufus Peebles, a Cambridge-based psychoanalyst who considers himself highly psychoanalytical, disagrees with the master. "Do by all means touch," says Dr. Peebles. "Sure, it evokes sexual feelings, but then they're out and we can talk about them. Touch, which talks to the subconscious directly, helps the patient get in touch with his early childhood experiences and feelings and thus grow faster in the here and now."

Touch's messages break down barriers, says Peebles, who literally holds, cradles, and nourishes his patients as though they were small babies.

"Some are uncomfortable at first," he acknowledges. "I suppose it does feel weird if you're over twenty to be held by a therapist, but it helps us to communicate in a powerful way." As babies, we find out who we are, maintains the psychotherapist, through touching our own bodies, and then we locate our parents through touch. "We see if they're there for us through our hands, and if you reach out a hundred times for your parents and they're not there, you'll stop reaching out."

Peebles calls his work "body-oriented psychotherapy," and he notes that hands often tell deeper truths than the dreams with which the traditional psychotherapist often works.

"When a need is felt, the body wants to make contact to express that need," says Peebles. "I need to grab your shoulder and find out if you're really there! It gives a person a feeling of competence when he or she *connects,* either in love or in anger."

Cese Mac Donald is a psychotherapist in New York who also respects the communicative value of touch.

"One day," she remembers, "a young woman called to tell me she'd been raped. Could she come over the next day to speak about it? *The next day???* I urged her to come immediately, and when she arrived, she was maddeningly calm and rational and understanding. I knew she was boiling inside—and words couldn't relieve her."

Mac Donald gave the young woman a life-sized male doll, put together with Velcro. "Pretend this is the attacker," she suggested. "I know you know the difference between a doll and a rapist, but just humor me and pretend for a while."

The young woman glared at the doll. She yelled epithets. Nothing. She felt self-conscious playing with a doll. Then something happened. She let go. She attacked the stuffed cloth figure, kicking and punching it, weeping, ripping off its arms. She connected. She touched. Her feeling could not be expressed in language because the rage she sustained was too terrible for words. It was touch that provided a place to put that fear and fury.

Our very first messages as infants, Dr. Peebles says, are in the form of touch; it's only later that communication becomes touch accompanied by words, and then, to our great loss, touch sometimes becomes totally supplanted by words.

Touch as communication surfaces in many ways. Men have more taboos than women about touching and being touched and, as a result, they find ways to pretend a touch is not really a touch. Many of these ways are called contact sports. Through acceptable games, men are allowed to express their feelings about competition, aggression, and even desires to hug others of their own sex. Macho engineers or truck drivers can feel the joy of a spontaneous embrace without being suspected or suspecting themselves of homosexuality. It's interesting to note the growth of contact-sport popularity in the female world. Perhaps as women make more incursions into traditionally male-dominated territory, their need to embrace being subverted to the handshake, the touch sports allow them to say more of what they used to be able to say in hugs. Even the spectators at a sports event take advantage of the spontaneous pats, embraces, kisses that are permitted among the same and opposite sexes who attend.

Other sports, some solo like swimming, which incorporates touch in a glorious primordial way, and some involving "object touch" like tennis or golf, allow pent-up furies to be "hit off" legitimately and give their devotees palpable, tactile pleasure.

Body language has been around for a long time but it wasn't until 1970 that an observant man named Julius Fast named and wrote a popular book about it. The book was immediately followed by dozens of other books, uncountable therapy groups, business seminars, and encounter sessions, which attempted to read thoughts and feelings from gestures. Eyes, mouth, sitting and standing postures, and, most of all, touch talked. If you were shy and inexpressive, even if you were, God forbid, massively retarded, you could still make your thoughts known with body language.

Talking was less accurate than touch. Fingers and body contact made liars out of orators.

It was as true a notion as any that ever came down the pike, which is why Mr. Fast attained fame and fortune for bringing the phenomenon to our intellectual attention.

The self-touching of young children often tells the world more than they want it to know. They have not yet learned to erect their body armor.

If a young child tells a whopping lie, she'll often cover her mouth with her hands.

If a young child doesn't love what his mother or playmate is saying, he'll spread his hands right over his ears.

If he sees a threatening sight, he covers his eyes. Three famous self-touching monkeys, See No Evil, Speak No Evil, Hear No Evil, did the same thing.

Adults are not immune to self-touching, note the experts, even though they're savvy enough to attempt a camouflage of the self-touches. Allan Pease, a management consultant in Sydney, Australia, notes that gestures are easily read if one's looking, and a person who is telling less than the truth will often guard his mouth as he "unknowingly instructs it to suppress the deceitful words that are being said." Interestingly enough, observes Mr. Pease, if a person objects to what someone else is saying, she'll often cover her own mouth as she listens, and one of the most disconcerting sights for a public speaker is to spot many of his audience covering their mouths in a way that says they think the speaker is a damned liar.

Other self-touches are the chin stroke, which also speaks worlds— something along the line of "s-u-r-e you can raise the money for that deal"; the nose touch, a thinly disguised version of the mouth cover, which tries to stop your mouth from lying, but, at the last moment, to be less obvious, heads for the nose. (Pease speculates that lying might cause the delicate nose nerve endings to tingle and that the nose touch assuages the tingle.) Eye and ear rubs are grownup versions of the child's wish to not see or not hear. The clenched-teeth gesture—touching your own teeth, as it were, is an attempt to prevent your own or others' unpleasant words, say experts. And scratching your neck, outer hand, cheek, or leg may be a sign of dry skin but more likely is a signal of doubt or uncertainty, says Pease, who also notes that most people scratch five times in the latter case. It is particularly noticeable when one's words contradict the uncertainty expressed by the scratching, as in "I know what you mean"—and you don't.

Ashley Montagu says that when one wishes to *be* touched, one often says it by touching oneself. Thumb sucking, for instance, is rarely seen at all in more primitive societies, where a baby is breast-fed whenever he feels like it and touching and hugging a child are ubiquitous.

Ever notice the rhythmic self-rocking and self-hugging in a nursing home or a mental institution? This is possibly due, say experts, to the touch deprivation that very old or very troubled people often experience. Indeed, autistic children rock themselves—especially blind autistic children; they seem to be trying to figure out where they end and the rest of the world begins by feeling the touch of the air moving over their skin. Head banging and twirling are especially common among the autistic.

The reassuring self-rocking movement is not the sole prerogative of the old or confused. Quite young and vigorous people have been observed resorting to this when they are greatly grieved. Also, self-rocking can be a matter of heritage. In my grandfather's orthodox synagogue, the Jews rocked and rocked and rocked in a mesmerizing dance as they prayed or studied or grieved. There is no question that the act is one that soothes and strengthens.

Dr. William C. Schutz in his book *Joy* has expressed the notion that those who self-touch in the form of tightly crossed arms and legs communicate resistance against anyone trying to touch them. It is possible through education, believes Schutz, to teach people to "unlock" their self-touching gestures and thereby unlock much of their obvious-to-all tightness and withdrawal. Whether or not education will release tension is debatable, but what is not even arguable is that a woman who shields her upper body with her tightly crossed arms, and her lower body with her clenched knees is not the most relaxed person at the party.

Leo Buscaglia, the best-selling advocate of love, also known as Dr. Hug, tells the story of the young woman in his class who sat in the back of the room, constantly rubbing her own arms, forever licking her own lips. A dog walked into his classroom and was instantly caressed by each person he approached with his wagging tail and trusting eyes. Suddenly, the young woman burst out crying, "I've been sitting here in desperate loneliness, wanting someone to see or touch me like that damned dog. . . . I could die here of loneliness . . . while you all pat that dog."

She was trying to send a message by touch. No one heard.

Think of the traditional villain in a soap opera who instantly gives himself away by rubbing his two hands together in glee. The gesture is repeated a million times a year as people who are not such villains rub their hands together in positive expectation, as little children and adults rub their hands together to express delight.

Think of the executive who leans way back in his chair, hands clasped behind his neck, expressing—what? You know. Absolute confidence.

Think of the communication of finger gestures—actually touching one's own fingers to each other. Prayer, irritation, satisfaction, or nervousness can all be expressed by fingering one's own fingers. One-finger messages are also eloquent: a thumbs-up gesture signaling that everything's okay, a thumbs-down gesture saying the opposite, a thumb-out gesture asking for a car ride, a middle-finger-up-the-rest-of-the-fingers-down gesture signaling "go to hell," or worse. Gestures differ from culture to culture but not so much that they're not instantly understandable in the context of the situation—wherever you are.

Think, finally, of the stress-related compulsions that tell so much about us: nail biting, obsessive hand scrubbing, hair stroking, cuticle picking—all self-touch, all expressive.

If touching one's self communicates, touching others positively blabs.

Perhaps there is no touching quite so complicated and full of meaning as the touching between lovers, but touch between other people is fraught with signals, as well. The words say, "Trust me," and the pat on the shoulder can say, "You're crazy if you trust me." The words say, "I care about your crisis," and the too-flip touch says, "Your crisis bores me silly." The words say "Yes" and the touch says "No, no, no!"

What kind of listening does a person have to do to get the real messages? Skin listening; that's the best kind.

The handshake—quite apart from its use as a power play—also says a lot about the shaker. A macho, too-powerful handshake tells you that the person behind the hand is concerned with his masculinity; his too-strong grasp is a dead giveaway of his fear of weakness. A limp, cold-fish handshake belies any protestation of warmth. It's also repulsive.

A double-handed shake—where someone grabs your hand in both of his—can speak disparate messages. If used by people who genuinely care for each other, the message is loving. If used by a "suspect"—someone you just met, a politician or an undeclared enemy—its message is insincerity or an attempt to control.

Ironically, some of our most public parts are our most private touching places. The head, for example. Even though the head and face are usually bare and open to public scrutiny, they're not often open to public touch. Experts have noted that people who easily touch each other on the arm or shoulder—or even thigh—are loath to show affection by face touching. To reach up and gently stroke the face of a friend with your hand is an extraordinarily personal and loving gesture. It speaks of trust and wanting to be trusted. It's rarely used.

Eye contact may not be a literal touch, but it is a powerful touch, nonetheless. Because it begins in earliest infancy between mother and infant, it carries a lifelong intimacy and potency. Looking deep into another's eyes reveals depths not touched by words. Social and other lies are often stripped bare by deep eye contact; the deepest trust between lovers is exchanged eye to eye.

Touch makes, if not intimates, then at least acquaintances out of strangers. It can also have quite the opposite effect. As sociologist Nancy Henly observed, touch is the nonverbal equivalent of calling someone by his first name. It reduces professional and social distances. It must be dealt with carefully. One can, however, overstep one's bounds with touch; always honest, touch gives away secrets. A failed attorney I know used to touch a prospective client as soon as one was ushered into his office. He meant to impart the message "Relax, you're safe here, we're friends," but the opposite message was given because he touched too eagerly, too soon. The client was "called by his first name" before they were friends, and the message received was "I'm desperate for clients and I'm not very sure of myself."

The state of the toucher is, of course, of the greatest significance. When someone tells you he's not afraid and therefore you shouldn't be either, his ice-cold hands will give you more angst than you ever could have managed alone. A nurse with rigid and stiff hands will not give sustenance to her charge; her touch will be less than hope-giving.

The communication that touch makes possible depends on the state of the person being touched as well as the state of the toucher. The same kind of touch on a sleepy baby and on a wide-awake, ready-for-trouble baby produces vastly different messages. To the sleepy toddler, the touch may croon, "Sleep peacefully," while to the baby about to step on the roller skate the touch may say, "Be careful."

The quality of a touch also determines its essential message. A perfunctory light shoulder tap may mean, "I'm here for you," but the same tap, in the same place, with a different, more urgent intensity, may say, "Don't mess with me."

Finally, the place you bear in the affections of the one who is being touched changes the message of a stroke. An arm around the shoulders of an equal may be loving and reassuring while an arm around the shoulders of a rival may take on positively malevolent connotations.

Touch is so potent a communicator, it's no wonder that we sometimes back off from it. The *lack* of touching sometimes says even more than the touching and it is usually a negative statement: "I am afraid of you or what's happening to you. If I don't touch you, I can keep my emotional distance."

Leonard Probst, the television drama critic and author, wrote a piece for the *New York Times* about his experience with cancer. "You are not quarantined or isolated by the disease," he wrote. "People are medically free to see, touch, hug and kiss you—and they do. That it is not catching is a great gift."

I wish I'd read that article when my own friend Diana was dying from the same disease eight years ago. I did all the right things, I thought, driving her to her chemotherapy treatments and listening to her fears. We talked and talked of every intimate thing, and finally she died and everybody said, "Oh, Sherry, what a good friend you are." But I knew it wasn't true. I couldn't do the one thing she needed most. I couldn't hold her, stroke her face when it was streaming with tears. I offered her Kleenex instead of touching. Years of being a grownup and keeping my physical distance from other women made me sit in the car, polite, proper, respectful of her agony. I should not have been respectful of her agony. I should have hugged her nausea away. Instead, my fear of her disease kept me at arm's length.

My touching of Diana would have communicated to her that I hated the thing that was killing her but she was the same person to me. All the listening and talking in the world didn't say that. Touch is not better than language. It *is* language.

THE WORD
(Touch *Becomes* Language)

Language is an abstract concept. Touch is more concrete than the Empire State Building. It's not by accident that a baby takes no time at all to learn to touch even though he takes a whole lot of time learning to talk. Before we have language, we reach out to touch objects because touching makes things real. We learn to explore our bodies, including our genital areas, with great interest, way before we have names for each part. Our early caretakers seem always to be barraging us with strange sounds that appear to relate to what we are touching:

- STICKY . . . don't touch
- HOT . . . don't touch
- SOFT . . . touch the pussy cat
- HEAVY . . . you can't lift it

Before we really have the foggiest notion of what all those sounds mean, we're touching everything we can get our hands on to get the feel of things. Then, and only then, do we match the word with the touch. Without a sense of touch, we might learn to speak, but I question whether we'd really develop the full ability to be comfortably fluent and knowledgeable about how the real world works. If you couldn't, for instance, *feel* an exhilarating ocean or lake, would you know what a swim really meant? Even if you learned to do forty-eight laps in twenty minutes? If you couldn't *feel* a kiss, would you love it? A person from the Caribbean may have read about snow, heard about it, seen pictures of it, but can he really know the meaning of snow unless he's *felt* a cold, wet snowball on his face, played in the soft drifts? Touch makes words and images concrete.

Helen Keller, who couldn't hear or see, said touch gave her language. Not only did it impart texture to her black and silent world, it actually gave meaning to the millions of bewildering ideas that cascaded about her bewildered body.

Dr. Alexander Lowen, in his book *Physical Dynamics of Character Structure*, wrote, "No words are so clear as the language of body expression once one has learned to read it." It's no wonder that touch, the ultimate tool of body expression, has lent its subtleties to the language of words. Touch words have seeped into our vocabulary and they are as explanatory as, well—a touch.

A generous person is a "soft touch." People can "rub you the wrong way," or "get under your skin," or be "touched in the head," or be "heavy-handed" or "thick-skinned" or "thin-skinned." Were you ever "struck" with an idea? Did anyone ever "stroke" your ego or have you ever carried a "touchstone"?

How many arguments have you "touched off"? And has your relationship with a difficult person ever been "touch and go"? Do you "touch up" your makeup? Do you have the "common touch"? Have you ever suffered so much you've "touched bottom"? No? Well, then, "touch wood" out of gratitude!

How many of us have been "touched" by kindness? Or touched by lust and felt the need to "scratch the itch"? We also tend to use words like "feel" in symbolic ways that have little to do with what feeling's all about. "I feel happy, I feel blue, I feel for you, I really do." "I felt as though a bomb fell in my bedroom." When we speak of emotions in terms of touch, it's almost as if we're saying that feelings like happiness, sadness, shock, or love can be actually touched in the same way one might touch a puppy or swatch of velvet.

It would seem that confusion might reign when one uses symbolic touch words for deeds or personality, but touch is so potent its precise meaning almost always comes shining through. If your natural language is English and a person has shortchanged you, he has definitely "stiffed" you. If you meet an uptight gentleman, you have surely met "a stiff." One touch word, two different meanings, absolute clarity. The cold, unpleasant stiff touch of a corpse makes you feel cheated of life. The sense of touch has made it all real.

I remember being a little girl and feeling exquisite pleasure when my cousin and I played the game of spelling out words on each other's back. Part of the pleasure, no doubt, came from the sensual pleasure of being touched, but a very real part of the enjoyment came from the tangible sense of *feeling* language. What a victory when we were able to decipher the back scratches! Researchers in Philadelphia, in Boston's Children's Hospital, and in other study centers around the world are working, right now, on alphabets that can be tickled out electronically onto the skin so that blind people will be able to read more efficiently than with Braille.

From a "touch of flu," which clearly undermines one, to a "touch of grace," which clearly uplifts one, to a "touch of class," which clearly marks one a favorite of the fates, touch as language clarifies our communication and adds color to our metaphors. Linguists as well as dilettantes in conversation have to be eternally grateful. Give me five!

CHAPTER EIGHT

FINISHING TOUCHES

A TOUCH OF SPOT

He was really terribly young to need home dialysis, but when his kidneys began to fail as part of the fallout from diabetes, thirty-one-year-old Randolph Curran was hooked—or hooked up, as the case might be. And he hated it. Formerly an athlete, Curran simply couldn't take the prison his body had become. He felt as if he were trapped in his own room; the hours spent tethered to the machine were agonizing to his psyche. He began to yell at his two kids almost constantly. His wife was ready to run away from home. Nothing tasted good, nothing was fun. He'd had it.

His wife, Allie, noted one day that he'd been stockpiling painkillers. He'd hidden his cache of pills in a shoebox under the bed, a sure sign to his wife that he was thinking about suicide. Frantic, she begged their friend, a psychiatrist, to make a house call for business—not for pleasure. And the psychiatrist confirmed Allie's fears. Yes, her husband was profoundly depressed. He had no wish to talk about anything except the old days, the good days. Oddly enough, mentioned the doctor, one of the best things about the old days, besides the tennis trophies, had been Curran's Dalmatian dog, who had died, along with Curran's good health. The dog had been his constant companion in his adolescence.

"Perhaps another dog?" the doctor tentatively suggested.

That was all she needed—a dog to walk and train. Didn't she have enough problems with a husband who relied on a machine for life and one who was increasingly difficult and sad, at that?

But something made her do it. She went out for yogurt one day and came back with an eight-week-old Dalmatian pup. The kids went berserk. They called the dog Spot so, Allie Curran says, she could yell, "Out out, damn Spot."

After the first few days of family fun, the kids went back to their friends and their fighting and their school.

Who didn't go back to his depression was Ralph Curran. He became absorbed in the puppy's welfare. He trained it, taught it tricks, walked it, and touched, touched, touched it. When Curran lay in his bed having his blood purified, Spot lay quiet, close, purifying his heart. Curran's hand rested for many hours a week on Spot's warm flanks. The man had a wife and children who loved him dearly, but it was the dog who gave him his will again; it was the dog who lifted the curtain of despair and regret; it was the dog who was almost always available for touch.

"Look—I hate my disease," Curran said in an interview recently, "but the disease isn't *me* anymore. I can divorce the tubes and the machines and the insulin from my spirit now. And you know why? It's that crazy dog, that insane, crazy animal that has somehow pulled me out of the ranks of the living dead. Don't ask me how."

Doctors don't know exactly how either, but there is a fast-expanding body of medical literature that is documenting the healing power of pets, not only in terms of lifting depression, but of calming troubled souls, lowering blood pressures, raising life expectancy and changing the way the health picture looks for millions.

At the University of Pennsylvania, there's an unusual group of hard-core researchers, gathered under the title of The Center for the Interaction of Animals and Society. The researchers include biologists, veterinarians, psychiatrists, and anthropologists. The center is partially financed by grants from such prestigious groups as the Dodge Foundation, the American Psychiatric Association, the Humane Society, and none other than the National Institute of Mental Health, all of which apparently believe in the Pet Factor. A member of the group is Dr. Aaron H. Katcher of the Department of Psychiatry at the University of Pennsylvania. Dr. Katcher's initial interest was the biological effects of loneliness on people who failed to incorporate other people into their lives. The effects were notable. In a 1977–78 study of patients at the University of Maryland Medical School Hospital, Katcher and his colleagues found that "people who are socially isolated have higher death rates and higher rates of serious illness than people who are married."

After the study, Katcher's group decided to take things a step in a direction none had earlier anticipated. They opted to include pets as a source of companionship.

"We tested the effect of animal companionship by looking at the one-year survival rate of patients who had been hospitalized for coronary heart disease. There were significantly fewer deaths among the patients with the pets," notes Dr. Katcher. Most surprising of all, the researchers also found that, even if the patients had enough human contact, the presence of a pet still worked wonders. "We concluded," says Katcher, "that pets had positive effects on health and that animal companionship was not necessarily a substitute for absent human associations but a unique kind of companionship of its own."

What physiologically happens with pet contact? The researchers put people and pets into experimental chambers where both were hooked up to monitoring machines. When a dog is petted by a friendly human being, there are large falls in monitored heart rate and blood pressure in both animal and person as each relaxes the other. Katcher and his group tested a horse and its trainer—the same result. Cats were an extremely potent pressure uncooker, apparently because of their strong audible and visible response to being stroked; a purring, kneading, rubbing cat works wonders.

Pet ownership, found Dr. Katcher and Erika Friedmann, a biologist and epidemiologist, accounted for a significantly higher survival rate in patients discharged from the coronary care unit.

In another study, Katcher, Friedmann, and Dr. James Lynch at the University of Maryland found that, when animal owners talked to other people, their blood pressure rose—sometimes dramatically—but, when they fondled their pets, their pressure dropped to below resting level.

Some foreign studies have been of great interest to Dr. Katcher and his colleagues. They include the work of a French veterinarian, Dr. Ange Corderet, who has provided the center with startling examples of severely disturbed and withdrawn children who, upon touching pets, have begun to reach out to people. A social worker named David Lee has documented the effects of owning small pets—like white mice and guinea pigs—for inmates at the Lima State Hospital for the Criminally Insane in Lima, Ohio. Usually violent patients seemed calmer, and one inmate, who went on a wild rampage through the hospital, destroyed everything in reach—except his pets, on whom he didn't lay a violent finger, just a loving one.

But it is with the elderly that pets may make the biggest difference. People who all their lives never looked twice at a cat or dog in old age find tremendous affinity with them. And as for the lifelong pet-lovers, well, their four-legged friends can be lifesavers. It doesn't matter what kind of an animal it is, as long as it's warm-blooded and you like it. Fish can never hack it. It's not the wag or purr or tweet that does it, either.

It's the way a cat has of bending in to you, purring with pleasure at your physical presence. It derives pleasure because you're there! That's a pretty comforting fact in itself. Very few people purr with pleasure at the presence of the elderly.

At the University of Maryland Hospital in Baltimore, 64 men and 28 women, almost all elderly, were hospitalized because of heart attacks or severe chest pain. All of them recovered after intensive treatment and eventually went home. One of the items noted in the hospital record was that, of the 92 patients, 39 did not own pets. Within a year of their hospital admissions, 11 of the patients who didn't have pets died. In contrast, all but 3 of the 53 pet owners were alive at the end of the year. The *Journal of the American Medical Association* notes that the findings "fly in the face of advice given by some physicians to dispose of pets because they might be an unnecessary burden."

Researchers began to investigate. In Britain a study was made of the effects of pets on social attitudes and mental and physical health of people aged 75–81. Some of the people who lived alone were given budgerigars—small, colorful, easy-to-care-for birds. Only some of these elderly owned TV sets as well. The investigators thought that a pet might matter less in the life of an older person who owned a TV set than of one who did not. The presence of the birds made a "consistent, significant difference in the lives of the study subjects, while the presence or absence of TV sets had no detectable effect." The pets became such a focal point for the elderly that their presence caused their owners to live in the present instead of the past. The new lives in their lives brought immediate responsibilities and pleasures, and many even forgot that they hurt.

In the U.S., humane societies across the country have reported spectacular success with the "prescription pet" for the elderly. In nursing homes, studies show that dogs served as "effective ice breakers and social matchmakers." Elderly residents, when encouraged to stroke an animal, began to relate to others for the first time. One spoke his first words in years, exclaiming in delight, "You brought that dog!" Use of warm-blooded pets doesn't cure old age, said one doctor, but it acts as a facilitator to bring people out of themselves.

"Pets may even help improve an older adult's diet and exercise patterns. If you are fixing dinner for the pet, you're more likely to fix something for yourself, as well. And pets need to be walked or lifted; one must bend down to brush or feed them.

Dr. Boris Levinson, the late New York psychiatrist, was a pioneer in what he called PFP—Pet Facilitated Psychotherapy. "An actual chemical change occurs in the brain when there is a pleasant interaction between the pet and the owner," observed Dr. Levinson. "Opiate-like chemicals are released which give the same sense of euphoria as certain drugs."

In short, an uncritical four-legged companion can be an instant Valium or, in contrast, an instant motivator, as Dr. Levinson found ten years ago, when an autistic child in his care said his first words to a tail-wagging dog.

Roger Caras, the famed writer of animal stories, tells of visiting an institution that treated confirmed alcoholics. A program was begun in which each new admission was given a puppy to raise at the hospital, and administrators found that touching and caring for the puppies created such a tight bond between troubled adult and animal that the adults learned to face responsibility at a level they could handle.

Marilyn Anderson is a volunteer worker at the Anti-Cruelty Center, a pet society in Chicago, Illinois, which offers a "pet on wheels" program. Regularly, the society brings animals into the hospitals and orphanages of the city to cheer the children.

"Touch is a pain reliever," she says. "I know it. I saw it. I'll never forget it. There was a little girl at the Shriners' Hospital for the Crippled who was scheduled for surgery because her legs were so misshapen she couldn't walk. Prior to the surgery, she loved to play with the puppies and the kittens. I came in one day to see if she'd had her operation to straighten her legs and she was in terrible pain. The doctors needed to withhold her pain medication for a reason I didn't understand. The child lay there moaning and wincing. The nurses were frantic with empathy but they couldn't help. I gave her a kitten. She held it and stroked it and stroked it. When she had a pain spasm, she stroked faster —not harder. The kitten loved it. And it gave her relief—I saw it."

Trish Turk, my typist, can write paeans to the soothing relief she gets from severe menstrual cramps when her cat curls up to sleep on her aching stomach. It's not the heat of the cat that does the trick, either, because heating pads don't have the same effect; it's the touch of a living pet.

Anderson tells of the autistic child she herself observed at Cook County Hospital in Chicago. No one ever got a response from her; she stared straight ahead and made eye contact with no one. Every day her mother came, and every day she stared. "One day, as other children around her played with a puppy I'd brought, she paid no attention, as usual, saying nothing. But, as I started to leave, she reached out and touched that puppy. My teeth almost dropped out! The whole place was in an uproar. 'She touched it,' her mother sobbed. 'She touched.' "

The healing of antagonism between people may also be effected through animal touch, thinks Dr. Katcher. When people touch animals, not only does blood pressure drop, but a "smoothing of the facial features" occurs, with a loss of signs of tension. Intimacy can be achieved with great rapidity, notes the doctor, and it can even be achieved in public—unlike with most forms of togetherness. In fact, suggests the doctor, perhaps those interested in studying closeness—and breaks in closeness—might consider studying people with "the presence of a dog in an experimental chamber, a veterinary clinic, or a home because this permits the study of intimacy without sex in the same way that the procedures of Masters and Johnson permitted the study of sex without intimacy."

Dogs are a means through which males, many of whom (in America at least) feel constrained from showing physical intimacy, can both express and receive affection in public, says Dr. Katcher. "It may well be that, for some men, the only outlet for an intimacy expressed through touch may be an animal."

There are some children—certainly included are adolescents—for whom touch is not easily given or received. Providing these young people with animals is a way of providing them with intimacy—and healing the holes dug by a lack of intimacy.

Washington State University College of Veterinary Medicine has established a People-Pet Partnership Council. One of its projects will be to promote horseback riding among organizations for the handicapped. Being in contact with a bigger, stronger creature—especially riding on top of it—makes you feel stronger, bigger, more powerful. Jennifer Breame, the chief physical therapist at the University of Pennsylvania Hospital, takes a group of twenty handicapped young people on a weekly ride. "These youngsters are used to looking up at the world from a wheelchair," she said, smiling. "On a horse's back, they can look down . . . a bond develops between horse and rider," a bond that makes the daily living with a handicap more bearable.

Dr. Michael McCulloch, a Portland, Oregon, psychiatrist, sometimes prescribes pets for a debilitated patient—to help him laugh. "If humor is determined to have a positive influence on illness and depression," says Dr. McCulloch, "then this may be one of the pet's most significant contributions to human health." It's true. Pets do make you laugh, as anyone who has ever seen a kitten chasing a ball or a puppy chasing its tail can attest. Pets may not be the "new messiah" of health, says Dr. McCulloch, but they surely seem to represent a new frontier for study.

Pet touch is becoming big business. There are even schools where animals are currently being trained as therapists. William Winter is director of Assistance Dogs International in New York, where dogs and cats are trained for the physically disabled, emotionally and mentally disturbed, multiply handicapped. "Anyone the Seeing Eye Dog people won't train for, such as the lame, the elderly, the very young, the balmy," Winter says, chuckling. It is an amazing form of therapy, he notes, and certainly the cheapest. What Winter calls "seeing heart dogs and cats" are touch repositories. Many people have a need for a pet simply because it's always available for intimacy and unself-conscious touch. We all need to be needed and we all melt to uncritical companionship. There are good reasons why millions of us endure scratched arms and couches; there *must* be reasons why we, in ungodly hours, wander the streets with pooper scoopers, scraping up unspeakable deposits.

The reason is instinctive. We seem to know, deep inside, that the touch and attention of furry pals heals body and soul.

TOUCH TABOOS

We've been communicating, healing and loving through touch for over a million years and it doesn't seem logical that we'd have put a whole set of limitations on this universally instinctive act. But, of course we have. Anything that good has got to be bad is the puritan ethic that motivates touch taboos.

Some touch taboos are cultural, others are shared by almost everyone, and most taboos are unspoken, unwritten, almost never discussed. How do people know what's allowed and what's not? They have to have very long antennae, which automatically retreat or expand depending on where they are and what they're doing to or with whom.

What can I touch? Whom can I touch? Can I touch every part of him or her? Can I touch every part of myself? Can I ever ignore a taboo?

"Taboo" comes from a Polynesian word that means the prohibition of an act. "Taboo" often applies to the dangerous or unclear, and in Polynesia the object of a taboo is believed to have a power so strong that he or it may be approached only by priests. To violate a Polynesian taboo might mean death to the offender; that is the only punishment strong enough to cleanse the community. Pretty strong words. Still, touch taboos are so powerful that even here today violating them makes for big trouble.

Taboos can work two ways: as prohibitions imposed by social custom they can put roadblocks in front of pleasure; but taboos created as protective devices can sometimes serve us quite well.

Whether they are useful or harmful, they remain ubiquitous. Touch taboos are as much a part of contemporary life as the stop and go signs on the corner. While the latter regulate traffic, touch taboos regulate relationships.

Touch taboos, most of them anyway, should be gravely considered. There's no doubt that some taboos have evolved intelligently; certainly one wants to discourage the intrusive touch of the subway groper, or the touch of the paternal boss figure who uses contact as a put-down. It is not in anyone's self-interest, however, to go through life tacitly denying oneself the pleasures and stimulations of frequent tactile contact. Thinking men will rethink the taboos that would have them use touch simply for sex and power. They will make concerted efforts to fight an upbringing that suggested friendly touch is for sissies, same-sex touch is the property of homosexuals, and inter-sex touch is best saved for bedroom or boardroom power plays. Thinking women will rethink these same taboos as well as any other taboo that would deny them equality in all things. Men and women, further, would do well to touch their children, touch their friends, touch themselves, and thus develop, to its finest, the forgotten sense.

Touch is so vital a need that no person should ever unwittingly adhere to all the taboos that shriek *"Don't touch!"* It is, too often, life-destroying advice.

Don't Touch Strangers

This has to rank among the most potent of taboos, and no wonder: touching strangers can provoke sex, violence, and space intrusions—all of which are closely associated with human contact. In a big city, touching strangers can *really* be dangerous—"give someone the wrong idea" —and so most of us don't. Strangers could be insane. Strangers could be dirty. Strangers could change our secure worlds if we invited them to intrude with a touch. It's safer not to touch. The problems come when we meet "okay" strangers in allowable situations—at parties, on blind dates, checking out museums or music. The don't-touch edict puts up

barriers in front of *nice* strangers as well as dangerous ones, barriers
that may take a long time to break down, when one friendly arm
squeeze may help pave the way to friendship. Still, most of us will not
break that don't-touch-strangers taboo. It would be a threat to the
society we've built if any old husband could spontaneously reach out
and touch any other old husband's wife—even if it were meant simply
as a friendly gesture. If a job applicant felt free to reach out and touch
the face of the prospective boss who's interviewing him, our cultural
structures would fall apart—to everyone's confusion.

The trouble with strangers touching is that we can't yet recognize
one another's touching codes. That makes us suspect one another.
When I meet someone and he grasps my arm significantly, what should
I read in the grasp? Does he want to go to bed with me? Talk to me?
Start a friendship? Hurt me in a subtle way? What does his touch
signify?

When you know the motive is sex, touch becomes even more taboo.
If, on some enchanted evening, you should meet a stranger across a
crowded room, then fly to his side—*but, don't touch him.* Even if you
are entranced with each other, and you want to reach out lovingly,
sexual touch upon newly meeting someone is *verboten.* And, with the
specter of AIDS hanging over us, going to bed with someone without
thoroughly checking him out—a taboo the previous generation
managed to break—is diminishing rapidly. Sexual contact with a stranger is becoming more taboo than ever.

There are exceptions to the don't-touch-strangers rule—people
who are allowed to touch strangers all the time, without others looking
at them strangely. A policeman can touch me if he's got a uniform on
and if he does it gently: I wouldn't mistake his custodial or directorial
touch for a threatening one. My religious leader can touch any stranger
he wishes to touch—he'll be allowed. Children can always touch anyone
without being a threat, even though their mothers sometimes have fits
when they do because their touching may put them at risk. People with
licenses to touch, like beauticians, doctors, masseurs, salespeople, can
touch strangers.

Most of us, though, like animals, must circle and sniff and look at
each other in a "checking-out" ritual that may last hours to years before
we reach out. It is a dance of small talk.

Without familiar-making small talk, touching strangers or being
touched by them has to remain taboo. *Groups* of strangers are even
more taboo than single strangers. We go to extreme lengths to avoid
touching one of a group and if in a crowded elevator or busy street, we

accidentally make contact, profuse apologies are in order: it is as if a casual touch is a slap. We hold in our bellies, make ourselves tiny, develop aches in our musculature—anything to avoid the stranger on the crowded bus.

If we're not allowed to touch strangers, we're also not allowed *not* to touch them once we've received permission for touch.

You're with your friend. You have packages in both arms. Your friend meets his friend and introduces you to that person who is to you a perfect stranger. But the simple act of introduction, which is a way of standing up for someone, saying "this guy's okay—you can trust me," puts a burden on you to touch the stranger if he offers his own touch. If your friend's friend offers his outstretched hand upon introduction, you will do any amount of clumsy shifting of packages in order to meet his touch with your own; anything else would be a direct affront, even if your shopping treasures tumble all about you from your efforts.

The dance floor is another arena where it's permissible to touch strangers and peculiar if you don't. At a party, an attractive stranger asks you to dance. The "don't-touch-strangers" taboo is lifted. Close contact is perfectly acceptable—while the music plays. Anyone who holds stiffly back is labeled odd or frigid. The music stops. Remove those hands from my person, if you please. It's very complicated.

Don't Touch Yourself

The oldest perhaps, and certainly one of the most powerful touch taboos is the one that tells you not to touch yourself *down there.* Many young people grow up without a clear idea of their own bodies' capacity for sexual pleasure because their explorations have to be furtive. Some don't even have a clear idea of the location of their erogenous zones. Religion has traditionally raised a disapproving eyebrow when it came to sexual pleasure, claiming that the body is the lowest part of the human being, and should be subservient always to the mind and soul. Sexual activity was to be enjoyed only if one had a sincere view toward procreation, and since self-touch never could do much in the procreation department, it has almost always been a sexual taboo in Western culture. You could go blind or crazy or worse, if you masturbated.

Some touch experts trace the self-touch taboo to the Biblical story of Onan, son of Judah (Genesis 38:9), who spilled his sperm on the ground when he interrupted coitus and masturbated instead of inseminating his brother's widow, as the Hebrew code of ethics directed. By not producing an heir for his brother, he fixed it so he'd inherit his dead brother's fortune and leave his sister-in-law poor. Not very nice, and masturbation was singled out as the culprit. (Many centuries later, the writer Dorothy Parker irreverently named her canary Onan because "he spilled his seed on the ground.") Today's experts who valiantly try to convince us that sexual self-touch is healthy have never completely succeeded in their efforts. Despite the fact that in flourishing modern sex shops all manner of creative aids to masturbation are openly sold, even the most ardent of our eighties population feel just a little guilty, just a little embarrassed to touch themselves, even in absolute privacy—a feeling which doesn't really *stop* them since presumably over 90 percent of all people engage in masturbation at some time or another.

Self-sexual touch isn't the end of it. We are so enjoined against self-touch that many other parts of our own bodies are forbidden zones, quite aside from those with sexual connotations. It's considered "disgusting" to touch the insides of the mouth, nose, or ear in public. We must keep our hands away not only from our breasts and genitals but also from our bottoms; presumably, it's unclean to linger lovingly on our own tushes, even though we're allowed to stroke our own arms or hair or faces or legs—below the knee, anyway. Children are still sometimes discouraged from sucking their thumbs even though current dental practice holds that thumbsucking is more comforting to psyche than dangerous to teeth and mouth.

It would be difficult to break the self-touch taboo frequently and still remain a viable member of society. Imagine the professor ruminating, while picking at his nose; imagine the attorney touching the velvet texture of his inner lip as he consults with a client. It doesn't play.

Don't Touch Your Same-Sex Friend

This taboo reigns stronger than ever—at least in the U.S. and Britain. Since gay men and women proclaim their gayness by touching those of their own sex, if you touch someone of your own sex, and you are not gay, you may still win a homosexual label—earned by a mere show of affection to a same-sex friend who is nothing more than a friend. Depending on where in the world you're touching, the taboo varies. In most European countries women may link hands and arms, walk with arms around each other's waists, but in much of America straight female touching female is taboo. Still, it's more acceptable than straight male touching male. Mothers may kiss grown daughters but, sadly, fathers don't often kiss grown sons. Mothers are allowed to caress their daughter's hair or body when sharing a talk, but if they try it with their sons, the sons usually withdraw, since they have heard of Oedipus. The taboo against explicit sex is so powerful that even some husbands and wives confine their touching to the bedroom, worried that their children will be aroused or confused. My own husband, one of the most touching and touchable men in town, is not thrilled about touching me in public, concerned that we'll seem to be "showing off" our love, concerned that bystanders will interpret his touch as purely sexual and not simply loving.

Don't Touch if You're Elderly or Ill

Sexual touch is so threatening to many that some hospitals institute No P.C. (No Physical Contact) rules. The medical establishment, which should be most aware of the extraordinary benefits that occur from touch, is often so turned off when people who are ill, retarded, or old express sexuality through contact that they forbid it. Elderly married couples are often cruelly separated in nursing homes. And any romance that might bud between elderly singles is an unseemly shocker.

In *Touching for Pleasure,* authors Adele P. Kennedy and Susan Dean describe a recent incident in a nursing home, where two of the senior citizens were missing at the dinner hour: "The alert spread quickly throughout the home and the search began. All the rooms and beds were empty and the outdoor patio bare. A nurse checked to see if any medication was missing from the storage closet and, upon opening the door, screamed. She found the man and the woman. They were embracing in silence. She quickly called for security. The two *sex offenders* were separated and escorted to their rooms. Families were called, conferences held, and doctors consulted. The consensus was that the two promiscuous culprits should not be allowed further contact. Humiliated and confused, frightened and guilt-ridden, the two rapidly withdrew from friends and family. Within weeks of the crime, they both died."

Don't Touch—Very Much—Your Opposite-Sex Pals

Another sexual touch taboo is that between opposite-sex friends, although many experts say that this taboo appears to be weakening. Touch between two heterosexuals is often seen as a sexual approach. Lingering over a hug with your best friend's husband, who is also your pal, can't bring you points from your best friend or your own husband.

Little pats and pecks must stand in for hearty hugs and kisses, even though you have nothing sexual in mind. Men and women force themselves not to hug overlong when they're not intimately connected in a sexual way; each fears the other might misinterpret. We tend not to reach up and touch faces in sympathy or empathy because—well, face touching seems too *personal*. Handshakes are accepted in greeting and leaving many friends, but the moment a single-handed handshake is implemented with a two-handed handshake, even friends tend to worry about sexual implications at the worst, or power implications at the least.

Stanley Jones of the University of Colorado, and his associate, Elaine Yarborough, studied sexual touch taboos among other touch taboos. "People often don't *know* what the code is," says Jones. "If they touch someone and are rejected, they may take it personally and stop trying."

Don't Touch for Power

Those who are savvy about the implications that touch carries in status place heavy taboos on certain power gestures, which may or may not be innocently given.

Researchers Jones and Yarborough found that many subjects reacted strongly against touches used to "rub in aggressive verbal statements" because they are, in fact, a double one-up. Combining a verbal put-down like "Putting on a few, I see" with a poke in the belly is asserting such verbal-tactile dominance that it physically assaults the one being put down. In sophisticated circles, there are definite taboos against touching people to move them out of your way or to a place you wish them to be. Treating humans as if they were chess pieces carries a silent taboo, thank God.

Researcher Nancy Henly points out that touch can often cancel itself out in power plays but that touch taboos have also created a nontactile society. She alludes to Desmond Morris's feeling that, "Our untouchability has to do with status, with not wishing to be touched by our inferiors and not daring to touch our superiors."

Social-status taboos are present within many societies and, although their reasons for being are often spurious, the taboo effects are extraordinarily effective. As I mentioned earlier, the caste system of India kept the untouchables away from those who consider themselves superior. In orthodox Jewish and other societies, a menstruating woman is taboo for sexual purposes because she is thought to be "unclean." Sick people, though not contagious, are quite taboo as a whole. Old people, as previously noted, are touch-tabooed by adults, but it is interesting to note that small children rarely pay this any heed.

Don't Touch Your Cousin

There are taboos against incest of any sort, so generous hugs, breast to breast, are rarely seen when relatives embrace. Instead, there are pecks on the cheek or sideways hugs, and rarely does one see opposite-sex relatives in uninhibited touch encounters. Certainly there are ancient incest taboos against child and parent sexual contacts, and this taboo often expresses itself in a withholding of touch contact between parents and children over toddler age. Many of us can recall creeping with delight onto our fathers' laps—and then suddenly laps were off base for no reason we could decipher and we began to feel faintly guilty that we'd ever climbed on the lap in the first place. Little ones become fettered with taboos when their reaching-out is curtailed. Many youngsters, say some experts, may provoke punishment out of a desire to be touched—even painfully touched. A touch that hurts is better than no touch at all. "Pain may then become valued as the means for physical contact and confirmation to the child that he or she does indeed exist," says Dr. James Hardison, dean of instruction at San Diego Community College and an expert in sociology. There should be little doubt that from this taboo spring the sado-masochistic behavior and sexual practices of some adults.

The need to touch remains omnipresent but the touch taboos weigh heavy on our hands. Better not to touch at all than break a taboo is the conclusion too often reached. We stop running our fingers through our mothers' hair and we don't kiss our fathers if we're their sons. The nice man at the paper stand remains untouched and we are even cautioned against the germs our friends bear. Uh, don't kiss your brother *quite* so much, dear; *don't* let that dog hump your leg; and stop immediately whatever you are doing to yourself under the covers.

There's only one thing to do and it takes bravery. The taboos are powerful but, although a few do protect, many of them are senseless.

First, judge the situation for yourself. If it seems safe, risk—and touch. Follow your instincts. Reach out.

TURN-OFFS AND TURN-ONS

It's never too late to stimulate an atrophied sense of touch—either your own or that of a loved one. Indeed, there are techniques that can help you hone not only your touching skills but your touch instinct. Masters and Johnson and their trained practitioners, for example, have taught thousands of people how to enhance their sex lives by developing their sense of touch. And holistic centers—such as the Open Center in New York, Interface in Boston, Esalen in California—give ongoing workshops in Yoga, Reiki, massage and many other arts that enhance general sensory awareness and enliven one's sense of touch. But you may not want or need all those instructors; there's a lot you can do by yourself.

Before I talk about touch turn-ons, however, I would like to mention the definite touch turn-offs—those things that block touch's pleasures.

The Chemical Turn-off

Stimulants or depressants that change sensory awareness usually change it for the worse. They turn off touch enjoyment. *Stimulants* include amphetamines ("speed" or "uppers"), cocaine and caffeine, and they increase the rate at which the heart beats, raise blood pressure, and sometimes make one feel more alert and confident. They can, however, also make one feel nervous, edgy, and irritable. Nervous, edgy, and irritable people are not famous for their touch sensitivity.

Because such chemicals also slow the circulation of blood to the skin, one often feels cold, as well as jumpy. Coldness doesn't make you feel like touching or being touched sensitively—although it may make you more receptive to being held for warmth. Whenever blood circulation is slowed, touch is not going to be so finely perceived. Moreover, many men, after taking stimulants, find it difficult to sustain erections or ejaculate, and so touch's potentially erotic effects are stymied.

Dr. Robert Scott, Medical Director of the Southern California Women's Medical Group and Assistant Professor of Obstetrics-Gynecology at the University of Southern California's Women's Hospital, says about amphetamines, "They drive your whole body at a pace that exhausts the reserves you need at all times," and, as such, can hardly be touch enhancers. One who is moving at a frenzied pace can not give or receive gentling touches. Amphetamines can stimulate the mind, but they jiggle up the body, producing jitters, and excitement reaches levels that are artificial and often dangerous. While cocaine may loosen up body inhibitions, it also depresses touch sensitivity.

Depressants include barbiturates ("downers"), narcotics, tranquilizers, and they *really* dull the touch sense. Muscle relaxers, tranquilizers and sleep inducers like quaaludes do break down inhibitions but tend to make you less sensitive to touch, says psychologist and touch connoisseur Dr. Russ Rueger. Valium, for instance, enhances the action of GABA, the brain chemical that lowers excitability levels, says Dr. Scott, and thus deadens touch reception.

Psychedelics are more complicated. They arouse the nervous system, which enhances touch receptivity, but they do it in a spurious fashion. Because they have great impact on the emotional centers of the brain, they distort images and turn fact into fantasy. One's touch receptors can't be trusted when one plays with LSD, mescaline, or similar psychedelic compounds. A hostile act can look and feel like a loving act —and vice versa. If there's one thing touch should be, it's trustworthy. How else could its effects be lasting and meaningful?

A sexual experience under the influence of LSD may be intense, may produce multiple orgasms, may make your lover's touch seem warm—but that's drugged magic. True feelings are subverted to potions that provide illusory feelings. Chemistry can fool touch, but the inevitable side effects or wake-up ultimately turns off instead of on.

The Alcohol Turn-off

"In small amounts," says Dr. Sheldon Kule, Unit Chief of the Department of Psychiatry at Long Island Jewish Medical Center in New York, "alcohol can be a relaxant but the crossover margin where it becomes a depressant or an anesthetic is slim," and so alcohol, like most drugs, interferes negatively with the fine tuning of touch. Alcohol can seem to break down touch barriers because it first stimulates the nervous system, but the moment one crosses Dr. Kule's "crossover margin," alcohol depresses the system and touch sensitivity.

The Tiredness Turn-off

Fatigue is a touch turn-off. When one is weary, she won't really hear the messages touch is reciting, she won't allow her tired body to respond to touch invitations or proclamations. Then what else but touch comes to the rescue! A gentle massage can stifle the turn-off, rejuvenate the touch receptors, and act as a turn-on in the face of fatigue.

The Weather Turn-off

Climate can dull touch receptors. Being in a too-hot, too-cold, too-damp, too-muggy atmosphere can jade touch sensitivity. It's as if, when the body has a surfeit of external temperature stimulation, it temporarily locks up its touch sensitivity.

The Too-Much-Sexual-Touch Turn-off

Sexual overstimulation can be a touch turn-off. Many sex manuals prescribe direct, continuous touch contact to the erotic parts of the body, such as clitoris, nipples, or earlobes, but touching or stroking these areas continuously only leads to numbness and then pain. Touch sensitivity can be enhanced by moving from sexual spot to sexual spot rather than concentrating on continuous touch contact in only one place.

The Too-Little-Touch Turn-off

Touch has to be cultivated. People often forget how good it can feel and what good it can do when they're out of practice. "Use it or lose it" applies to everyday touch as well as sexual touch. Many people learn to activate touch receptors by paying a whole lot of attention to their skin surfaces. They apply creams and lotions, they ask to be tickled or scratched gently, they wear clothing that is pleasurably stimulating rather than constricting; they salute their own "touchability" every day of their lives.

And then there are the touch turn-ons. They are everywhere. All you have to do is look . . . and touch.

The Chemical Turn-on

Although most drugs act as turn-offs, one—whether the mothers of America want to hear it or not—acts as a turn-on—with a catch. Dr. Sheldon Kule, among other experts, says that "the only drug that can enhance touching—and thus sexual pleasure—is marijuana." The drug does heighten sensory awareness in a sort of mixed-bag effect—a little stimulation, a little depression, and even, say some experts, a little bit of hallucinogenic effect. "However," cautions Dr. Kule, "if your mind-set *before* you smoke grass says that the drug will act as a touch enhancer or an aphrodisiac, it will. If you're in a good mood, your touch receptors will be stimulated positively. *But,* if you feel angry or depressed before you smoke pot, you will be turned off to touch. The drug will exaggerate your down mood. The marijuana effect depends entirely on what you feel before you smoke."

Moderate users of pot report that textures seem more varied and intricate, soft objects more desirable to touch and hold. Massage and cuddling seem, they say, more pleasurable. Naturally, the working word here is moderate. Heavy usage, report experts, nullifies the touch-enhancing effect and simply creates zombies.

The Alcohol Turn-on

We've said that alcohol in larger amounts is a touch-off. A *bit* of wine, a glass of champagne, a can of beer or a shot of scotch can heighten touch awareness because it relaxes. One little bit too much and the alcohol starts depressing the nervous system.

The Aquatic Turn-ons

Warm baths gently touch receptors. Since the mind-body connection is a real thing, when the body is soothed and relaxed and ready for touch, the mind is equally jogged into touch appreciation. Bath accessories—loofah sponges, bath oils, back brushes, all incite touch sensibility and leave one on the receptive edge of sensation.

The Product Turn-ons

In today's "everything's for sale" marketplace, many products are available as touch enhancers. Some work on the acupressure points of the body, those places where energy blocks can be unblocked; some roll, some massage, some heat, some scrub, some whirl, some spray, some make you feel as though you're floating, some make you feel as though you're boating, some vibrate you sexually, some cuddle you motheringly—all enhance touch and make you more aware of what pleasurable skin contact can mean.

Specifically, you might try:

- A thermal heat massager (you lie on it and it warmly massages your back)
- A wooden body roller that loosens tight muscles and energizes the skin surface (try specialty shops like Abercrombie & Fitch or Hammacher Schlemmer)
- Wooden foot massagers available in health and department stores; the round, dumbbell-shaped roller sits on the floor and your feet rest on the bar between the two dumbbells as the rolling massage motion revs up circulation and evicts fatigue from tired feet!
- A back scratcher (any city's Chinatown carries this little item)
- Hot tubs, whirlpools
- Vibrators: Not only used for sexual purposes but wonderful for massaging every body part, heads, backs, muscles, and aching joints.
- Shower attachments that spray, "needle," massage, pulse, steam, stream—you name it. A stellar sensory plaything.
- Water beds: Some people claim that every sense is enhanced by floating all night.
- Makeup and skin-care products: Anyone who disdains the lure of the creams, lotions, perfumes, clay, mud, herbal masques, is missing a good bet. They promote, in addition to skin-touch stimulation, cleanliness, clear skin, conditioning, and beauty. Think about how a facial feels: the cool creams soaking deep, the hot towel as contrast, the soothing facial massage, the skin buffers to clear off the dead skin cells —all titillation to the sense of touch.

The Environment Turn-ons

Nothing turns on touch more than awareness of one's surroundings. Experience the texture of a gnarled tree, as you pass. Touch a petal of your neighbor's roses. When you get hit with a rain shower, don't huddle inward—extend your face like a flower and feel the sweet wetness. Think about the clothes you buy; never put anything on your body that itches—no matter how chic. Always try to go for the natural fabrics, the silks, cottons, velvets, soft wools, angoras that feel fine to the touch. If you like puppies, never pass one without stroking it. Sense the cool smoothness of an ice-cream cone before you even taste it. Luxuriate in the warm sun—feel it melt your tension.

The People Turn-ons

At a time in human history when it is most important for people to make contact, they seem to be drawing farther apart. Nothing turns on love among families, generations, or countries more than simple touch. Metaphorically and practically speaking, if men and women don't walk hand in hand through life, there is very little hope for love.

People touch can be learned. Even if you've never kissed your best friend directly on her skin, but instead kissed the air around her, pecked her cheek, touched her arm—kiss her today, directly, bluntly on her facial skin. Sense how that face feels under your lips. Linger a minute. Savor the feeling. You're allowed. No kidding.

Touch your baby. *Constantly*.

If your adolescent shies from you in typical adolescent rebuff, don't let him scare you off. Rub your hand along his arm. Kiss the place where his new beard sprouts.

When you read to your toddler, hold him in your arms. Bathe him with your touch as well as your tale.

Hug your mom. *Do it!* Allow yourself to enjoy it. Try not to think about past battles with your parents. Hug and hug.

Hold hands with your lover or your wife in the street. Get used to touching all the time—not just in bed. If she moves away, try again. Never let a movie run its course without hand touches with your partner.

Touch your doctor, your lawyer, your editor, your typist. The salesperson. Touch warmly, gently, appropriately (which means never touch the Queen, poor thing).

If you know an old person, touch him so much. Remind him of life. Age often makes us feel fragile and isolated. Touch helps.

Ultimately, the skin is not a cloak that simply covers you up; it is a channel to your essential heart. Receive touch with pleasure. It is a rare gift.

Give it back.

RESOURCES

Each of the organizations listed below is presented in the spirit of research and inquiry to make the reader's quest for information easier. None is recommended by the author. There are poor practitioners in each of life's disciplines, and in the final analysis one must use one's own experience and judgment to determine what works—and what doesn't.

A general method for getting information about touch therapies and practitioners:

- First, check with the information center of a large teaching hospital in your area.
- Second, try looking up the particular therapy under its alphabetical listing in the white pages of the telephone directory.
- Third, in the yellow pages, try listings under:
  - Physicians
  - Pain Control
  - Holistic Health Centers (or Holistic Health)
  - Alternative Medicines
  - Japanese (or Chinese or whatever ethnic society, consulate, bureau might be helpful)
- Fourth, try asking your own physician or the local health store for a recommendation.
- Fifth, your local library is an invaluable source. Find books about holistic health, which will invariably list centers and organizations. These will then recommend specific practitioners. Other books are a wealth of information about massage, parent-infant bonding, etc.

ORGANIZATIONS FOR GENERAL INFORMATION

The following are valuable sources of information if a call to an organization for a specific touch technique is unproductive, calling a large, general organization can often help you find a resource.

American Holistic Health Sciences Association
1766 Cumberland Green, Suite 208
St. Charles, IL 60174
(312) 377–1929

American Holistic Medical Association
2727 Fairview Avenue East
Seattle, WA 98102
(206) 322–6842

Association of Holistic Health
P.O. Box 9352
San Diego, CA 92109
(619) 425–0618

The Association for Research and Enlightenment (psychic healing)
(ARE Founded by Edgar Cayce, 1931)
P.O. Box 595,
Virginia Beach, VA 23451
(804) 428–3588

Healing Light Center (psychic healing research and training)
204 East Wilson
Glendale, CA 91206
(213) 244–8607

Interface
230 Central Street
Newton, MA 02166

National Self-Help Clearing House
(A group to point you in the right direction about almost any touch therapy)
1012 Eighth Avenue
Brooklyn, NY 11215
(718) 788–8787

Omega Institute
Lake Drive, R.D. 2
Box 377
Rhinebeck, NY 12572
(914) 338–6030

The New York Open Center
83 Spring Street
New York, NY 10012
(212) 219–2527

Whole Life Magazine ($20.00 for 8 issues; alternative health therapists advertise)
89 Fifth Avenue
Suite 600
New York, NY 10003
(212) 741-7274

ORGANIZATIONS
FOR SPECIFIC TOUCH TECHNIQUES

Birthtouch (p. 20) and Touch and Babies (p. 23)

Any obstetrician is able to recommend a favorite Lamaze group or other childbirth instructor. Most hospitals give courses as well. Contact the following organizations for more information about classes and coaches in your area:

American Society for Psychoprophylaxis in Obstetrics (Lamaze)
1370 Lexington Avenue
New York, NY 10028
(212) 831-9327

National Association of Parents and Professionals for Safe Alternatives in Childbirth
P.O. Box 267
Marble Hill, MO 63764
(314) 238-2010

Books: Frederick Leboyer, *Birth Without Violence.* Knopf, 1975.
         Frederick Leboyer, *Loving Hands.* Knopf, 1976.
         Marshall Klaus, *The Amazing Newborn.* Addison-Wesley 1986.
         Joseph Chilton Pearce, *Magical Child.* Dutton, 1977.

Therapeutic Touch (p. 66)

There are many Therapeutic Touch teams in major teaching hospitals in major cities across the country—in Boston, New York City, Austin, Tucson, San Francisco, and Portland, Oregon, to name a few. Call the large hospital in your community (the nursing department might be a good place to start) to ask if Therapeutic Touch is available. If you have no luck, a call to Dolores Krieger's home territory will bring suggestions:

Nurse Healers-Professional Associates, Inc.
175 Fifth Avenue, Suite 3399
New York, NY 10010
(212) 838-8083

Pumpkin Hollow Farm Foundation
Box #135, R.R. #1
Craryville, NY 12521
(518) 325-3583

Acupuncture and Acupressure (p. 75)

Center for Chinese Medicine
2303 South Garfield Avenue, Suite 202
Monterey Park, CA 91754
(213) 573–4141
(Ask for a list of member practitioners across the country specializing in acupressure and acupuncture)

Chinese Massage (p. 83)

Center for Chinese Medicine
2303 South Garfield Avenue, Suite 202
Monterey Park, CA 91754
(213) 573–4141

Shiatsu (p.84)

Again, start with the information department of your local teaching hospital. Then try the white pages under Shiatsu, and the Yellow Pages under *Wholistic* or *Holistic Medicine, Alternative Health Therapies.* Ethnic groups like the Japan Society, the Japanese Convention Bureau or the Japanese consulate nearest you might have recommendations. The best way to find the most legitimate practitioners is usually through word of mouth: a friend, a physician, or another practitioner of Eastern medicine.

Reiki (p. 87)

The Reiki Alliance
4114 East Huntington Boulevard
Fresno, CA 93702

The American International Reiki Association
P.O. Box 86038
St. Petersburg, FL 33738
(813) 347–3454

Polarity Therapy (p. 91)

Polarity Wellness Center of New York
38 West 28th Street
New York, NY 10001
(212) 889–3555

Reichian Massage (p. 93)

The American College of Orgonomy
P.O. Box 490
Princeton, NJ 08542
(609) 821–1144

The Institute for Bioenergetic Analysis
144 East 36th Street
New York, NY 10016
(212) 686–2844

Bioenergetics (p. 96)

The Institute for Bioenergetic Analysis
144 East 36th Street
New York, NY 10016
(212) 532–7742

Rolfing (p. 100)

Rolf Institute
Box 1868
Boulder, CO 80302
(303) 449–5903

The Alexander Technique (p. 102)

The Alexander Technique
142 West End Avenue
New York, NY 10023
(212) 799–0468

The Feldenkrais Technique (p. 104)

Feldenkrais Guild
P.O. Box 11145
San Francisco, CA 94101
(415) 550–8708
(The guild has a list of all recognized therapists.)

Osteopathy (p. 107)

American Osteopathic Association
212 East Ohio Street
Chicago, IL 60611
(312) 944–2713

Chiropractic (p. 110)

American Chiropractic Association
1916 Wilson Boulevard, Suite 300
Arlington, VA 22201

Applied Kinesiology (p. 113)

Same as for Chiropractic

Touch for Health (p. 118)

Touch for Health Foundation
1174 North Lake Avenue
Pasadena, CA 91104
(213) 794–1181

Swedish Massage (p. 121)

Many people are puzzled about how to find a good massage without falling into the grips of a "massage parlor." Licenses in many states as well as registration are required but this means little since almost anyone who puts in the hours and money at a state-approved massage school (there are myriads of them) can receive a license, regardless of skill. Often the best masseurs are men or women who have recently immigrated from Russia or Sweden or elsewhere with no language or public-relations skills—just golden hands. The only way to tell is to try.

Health spas (the local YMCA included) almost always have skilled masseurs on staff, and in most such places, one needn't be a member to buy a massage. Many practitioners are more than willing to give you their cards for future massages at their homes (sometimes less costly) or even yours. It is *not* an inspired idea to find a masseur in the back pages of a "fringe" newspaper or even an avant-garde community press; the masseurs who advertise might well be of the "sex-massage" genre and you may be in for more than you planned. Word of mouth is, of course, the best.

Reflexology (p. 123)

International Institute of Reflexology
Box 12462
St. Petersburg, FL 33733
(813) 343–4811
(Ask for a list of qualified practitioners)

BIBLIOGRAPHY

BOOKS

Allon, Natalie. *Urban Life Styles.* Dubuque: William C. Brown Co., 1979.

Althouse, Lawrence W. *Rediscovering the Art of Healing.* Nashville: Abingdon Press, 1977.

Atkinson, D. T., M.D. *Magic, Myth and Medicine.* Evanston, Ill.: World Publishing Co., 1956.

Benson, Herbert, M.D. *The Mind/Body Effect.* New York: Berkley Publishing Corp., 1980.

Borelli, Marianne D., M.A., R.N., and Heidt, Patricia, Ph.D., R.N., eds. *Therapeutic Touch.* New York: Springer Publishing Co., 1981.

Brown, Catherine Caldwell, ed. *The Many Facets of Touch.* Skillman, N.J.: Johnson & Johnson Child Development, 1984.

Eisenberg, David, M.D. *Encounters with Qi (Exploring Chinese Medicine).* New York: W. W. Norton and Co., 1985.

Fast, Julius. *Body Language.* London: Sheldon Press, 1984.

Goldstein, Bernard. *Introduction to Human Sexuality.* New York: McGraw-Hill Book Co., 1976.

Grossman, Richard. *The Other Medicines (An Invitation to Understanding and Using Them for Health and Healing).* Garden City, N.Y.: Doubleday & Co., 1985.

Hall, Edward T. *The Hidden Dimension.* Garden City, N.Y.: Doubleday & Co., 1966.

Hardison, James. *Let's Touch.* Englewood Cliffs, N.J.: Prentice-Hall, Inc., 1980.

Hastings, Arthur C., Ph.D., Fadiman, James, Ph.D., and Gordon, James S., M.D., eds. *Health for the Whole Person (The Complete Guide to Holistic Medicine).* Boulder, Colo.: Westview Press, 1981.

Henly, Nancy M. *Body Politics (Power, Sex and Non-Verbal Communication).* Englewood Cliffs, N.J.: Prentice-Hall, Inc., 1977.

Inglis, Brian, and West, Ruth. *The Alternative Health Guide.* New York: Alfred A. Knopf, 1983.

Inkeles, Gordon. *The New Massage.* New York: Perigee Books, 1980.

Kennedy, Adele P., and Dean, Susan, Ph.D., *Touching for Pleasure.* Chatsworth, Cal.: Chatsworth Press, 1986.

Klaus, Marshall H., M.D., and Kennell, John H., M.D. *Parent–Infant Bonding.* St. Louis, Mo.: C. V. Mosby Co., 1982.

Krieger, Dolores, Ph.D., R.N. *The Therapeutic Touch (How to Use Your Hands to Help or Heal).* Englewood Cliffs, N.J.: Prentice-Hall, Inc., 1979.

Lauder, Estée. *Estée.* New York: Villard Books, Random House, 1985.

Lee, Linda, and Charlton, James. *The Hand Book.* Englewood Cliffs, N.J.: Prentice-Hall, Inc., 1980.

LeShan, Lawrence. *The Medium, the Mystic and the Physicist.* New York: Ballantine Books, 1974.

Locke, Steven, M.D., and Colligan, Douglas. *The Healer Within.* New York: E. P. Dutton, 1986.

Mayo, Claro, and Henly, Nancy M., eds. *Gender and Non-Verbal Behavior.* New York: Springer-Verlag, 1981.

Meyers, Jeffrey, ed. *D. H. Lawrence and Tradition.* Boston: University of Massachusetts Press, 1984.

Montagu, Ashley. *Touching (The Human Significance of the Skin).* New York: Harper & Row, 1978.

Morris, Desmond. *The Naked Ape.* New York: Laurel Edition of Dell Publications, 1984.

Pease, Allan. *Body Language.* London: Sheldon Press, 1984.

Rueger, Russ A. *The Joy of Touch.* New York: Wallaby Books, 1981.

Samuels, Michael, M.D., and Bennett, Harold. *The Well Body Book.* New York: Random House/Bookworks, 1973.

Siegel, Bernie S., M.D. *Love, Medicine and Miracles.* New York: Harper & Row, 1986.

Simonton, Stephanie Matthews. *The Healing Family (The Simonton Approach for Families Facing Illness).* New York: Bantam Books, 1984.

Smith, Anthony. *The Body.* New York: Avon Books, 1968.

Thie, John F. *Touch for Health.* Romford, U.K.: D. D. De Vorss and Co., 1979.

Turnbull, Colin M. *The Human Cycle.* New York: Simon & Schuster, 1983.

Verny, Thomas, M.D., and Kelley, John. *The Secret Life of the Unborn Child.* New York: Summit Books, 1981.

Weil, Andrew, M.D. *Health and Healing (Understanding Conventional and Alternative Medicine).* Boston: Houghton Mifflin Co., 1983.

JOURNALS AND MAGAZINES AND UNPUBLISHED PAPERS

Crusco, April, and Wetzel, Christopher G. "Touch." *Personality and Social Psychology Journal* 10, no. 4.

Dowling, St. John, MB MRCGP. "Lourdes Cures and Their Medical Assessment." *Journal of the Royal Society of Medicine* 77 (August 1984).

"Electricity and Bone Healing." *Harvard Medical School Health Letter* (October 1981).

Erikson, Joan Mowat. "Vital Senses: Sources of Lifelong Learning." *Journal of Education* 167, no. 3 (1985).

Field, Tiffany. "Touching Newborn Infants." *Science News* 127, no. 301 (1984).

Frith, Greg H., and Lindsey, Jimmy D. "Counselors, Handicapped Students and the Touch Domain: A Recapitulation." *School Counselor* 31 (September 1983).

Grad, Bernard. "Some Biological Effects of the 'Laying on of Hands': A Review of Experiments with Animals and Plants." *Journal of the American Society for Psychical Research* 59, no. 2 (April 1965).

Harris, T. George. "Healers in the Mainstream." *American Health* (May 1984).

Heslin, Richard. "Steps Towards a Taxonomy of Touching." Paper given at Midwestern Psychological Association, May 1974.

Heslin, Richard, Whittier, Tommy, and Abellow, Rodolfo. "When Salespersons Touch Customers." Account of experiment at Purdue University, unpublished.

Jourard, S. M. "An Exploratory Study of Body Accessibility." *British Journal of Social and Clinical Psychology* 5 (1966).

Krieger, Dolores. "Healing by the Laying on of Hands as a Facilitator of Bioenergetic Exchange: The Response of In-Vivo Human Hemoglobin." *International Journal For Psychoenergetic Systems* 2 (1976).

Lynch, James, Ph.D., et al. "Effects of Human Contact on the Heart Activity of Curarized Patients in a Shock Trauma Unit." *American Heart Journal* 88, no. 2 (August 1974).

Lynch, James, Ph.D., et al. "The Effects of Human Contact on Cardiac Arrhythmia in Coronary Care Patients." *Journal of Nervous and Mental Disease* 158, no. 2 (1974).

Major, Brenda, and Heslin, Richard. "Perceptions of Cross-Sex and Same-Sex Non-Reciprocal Touch: It Is Better to Give Than to Receive." *Journal of Non Verbal Behavior* 6 (1982).

Older, Jules. "Teaching Touch at Medical School." *Journal of the American Medical Association* 252 (1984).

Robb, Christina. "The Healing Touch." *Boston Globe Magazine,* 15 November 1981.

Sokolow-Molinari, Ellen. "The Gift of Reiki." Unpublished paper.

INDEX

Acupressure vessels, 119
Acupuncture (and acupressure), 75–83
  anesthesia use of, 78
  information resources on, 190
  origins of, 75–76
  recommended uses of, 82–83
  sensations during session of, 80–81
  theory of, 77
  Therapeutic Touch compared to, 70
  Western interest in, 78–80
Adolescents, 37–39
Aging. *See* Elderly
Ainsworth, Mary D. Salter, 33
Alcohol consumption, 181, 183
Alcoholism, pet touch and, 170
Alexander, F. Mathias, 102–4
Alexander technique, 102–4, 192
Allon, Natalie, 134, 149
Als, Heidelise, 27–28
Amazing Newborn, The (Klaus), 32
Amphetamines, 180–81
Anderson, Marilyn, 170
Anesthesia, acupuncture used for, 78
Animals. *See also* Pet touch
  dolphins, 156
  mammal babies of, 25, 28–29
  primates, 8–10
Anthropology, 8–19
  ancient healing traditions and, 10
  cultural differences and, 16–19
  Eskimos and, 11
  massage and, 30–31
  neolithic drawings and, 10
  primate research and, 8–10
  sex and, 46–47
  tribal customs and, 11–14
  Western customs and, 14–16
Apnea, in babies, 28
Applied Kinesiology, 113–17
  information resources on, 192
  recommended uses of, 117

Argas, Sandy, 88
Arigo, Ze, 136–37
Atimelang tribe, 14
Auckett, Amelia, 30–33
Autism
  pets and, 170–71
  self-touching and, 159–60
Awareness Through Movement, 105–6
Azande tribe, 12

Babies, 23–24
  birth defects in, 29
  bonding between parents and, 31
  infant mortality, 23–25
  intellectual development of, 31–32
  massaging, 30–33
  premature, 23, 27–29
  primate experiments and, 9
  swaddling, 13
  touch and communication with, 26
  touch healing, 130, 131
Babylonia, touch healing in, 57–58
Baby Massage (Auckett), 30
Bali, 12
Bathing, touch enhancement and, 183
Bengssten, Otelia, 72
Benson, Herbert, 139–41
Bible, faith healing described in, 135
Bing, Elisabeth, 20
Biochemical studies, Therapeutic Touch
  research and, 72
Bioenergetics, 96–97
  information resources on, 191
Bioenergetics (Lowen), 96–97
Birth, 1–2, 20–23
Birth defects, 29
Blacks, traditions of, 14–15
Body language, 158–59
Bonding between baby and parents, 31
Bones, touch healing and, 132
Book of Common Prayer, 59

Brain, sex and, 42
Brazelton, T. Berry, 130
Breame, Jennifer, 171
Bresler, David, 79
Buddhism, Reiki and, 88
Burn victims, 29–30
  touch healing and, 129
Burwell, Sidney, 142
Buscaglia, Leo, 160
Butt patting, 16

Caesarean disorientation, 21
Cancer
  communicative touching and, 162–63
  faith healing and, 137
  touch healing and, 128, 131
Caras, Roger, 170
Carnegie, Dale, 153
Cartesian dualism, 64
Cayce, Edgar, 136
Center for the Interaction of Animals
  and Society, 167
Cerebral palsy, touch healing and,
  131–32
Chakras (energy centers), 69
Chapin, Henry Dwight, 24
Character Analysis (Reich), 94
Charcot, Jean Martin, 61
Charles II, King of England, 60
Charlton, James, 17
Chemicals. See Drugs
Chi. See Qi
Children, 34–39
  body language of, 158–59
  dying and, 53–54
  handshakes of, 145
  idle touching and, 35
  persuasive touching and, 154–55
  pet touch and, 171
  puberty and, 35–36
  somatosensory deprivation and, 36–37
Chimpanzees, 9–10
China
  acupuncture in, 76–79
  customs of, 18
  medicine of, 63–66
Chinese massage, 83–84
  information resources on, 190
Chiropractic, 110–13
  applied kinesiology and, 114, 117
  definition of, 110
  information resources on, 192
  office visit described, 111–12
  osteopathy compared to, 108–9
  recommended uses of, 113
Christian church, touch healing and, 59
Chusid, Emanuel, 29
Comer, James P., 14
Communication, 5

Communicative touching, 156–63
  body language and, 158
  language and, 163–65
  psychotherapy and, 156–57
  self-touching and, 158–61
  touching others and, 161–63
Corderet, Ange, 168
Coronary-care patients
  faith healing and, 138
  pet touch and, 168–69
  touch healing and, 129–30
Crowds, 16–17
Crusco, April H., 155

Dean, Susan, 40, 177
Depressants, 181
Descartes, René, 64
Dolphins, 156
Downing, George, 91
Down's syndrome, 29
Drugs, 180–83
Dying, 13–14, 39, 53–56
  need for touch and, 53–55
  rural customs and, 14
  therapeutic touch and, 55–56

Ebers Papyrus, 57
Eczema, touch healing and, 131
Egypt (ancient), touch healing and, 57
Eisenberg, David
  acupuncture and, 76–80
  Chinese massage and, 83–84
Elderly, 49–53
  anxiety over touching and, 49–50
  pet touch and, 168–69
  reduced opportunity to be touched,
    50–52
  self-touching and, 159
  sex and, 52, 177
  wheelchairs and, 51–52
Electromagnetic fields
  polarity therapy and, 91
  Therapeutic Touch and, 74
Encounters with Qi (Eisenberg), 76
Endorphins
  acupuncture and, 79
  Therapeutic Touch and, 74
Energy medicine, 63–97. See also specific
    therapies
  acupuncture (and acupressure), 75–83
  bioenergetics, 96–97
  centers in body, 92
  Chinese massage, 83–84
  Eastern medicine and, 63–66
  polarity therapy, 91–92
  Reichian massage, 93–97
  Reiki, 87–90
  Shiatsu, 84–87
  Therapeutic Touch, 66–75

England
  spiritual healing practices in, 141
  touch healing and, 59–60
Episcopal Church, laying-on of hands
    re-established, 59
Erikson, Joan Mowat, 51–52
Eskimos, 11
Estebany, Oskar, 71–73
Evans, Laurie, 33
Extrasensory perception, 140
Eye contact, as a form of touching,
    161–62

Faith healing, 135–42. *See also* Laying-on
    of hands
  cures credited to, 136–38
  medical establishment and, 138–41
  psychosomatic illness and, 138–39,
    141–42
Fanslow, Cathleen A., 52, 55–56
Fast, Julius, 158
Fathers, birthing process and, 20–23
Fatigue, 181–82
Feet, reflexology massage and, 123–26
Feldenkrais, Moshe, 104–7
Feldenkrais technique, 104–7
  information resources on, 192
Fetzer Institute, 65
Field, Tiffany, 23
Fisher, Jeffrey D., 129
Food allergies
  applied kinesiology and, 116
  Touch for Health and, 120
Football, 16
Forbes, Allen, 15–16
Foundling hospitals, infant mortality in,
    23–25
France
  customs of, 17–18
  touch healing and, 60–61
Frank, Jerome D., 140
Freud, Sigmund, 93, 156
Friedmann, Erika, 168
Friends, taboos against touching, 177–78
Functional Integration, 105
Furumoto, Phyllis Lei, 89

Galen, 59
Geriatric. *See* Elderly
Germany, touch therapy for babies in, 24
Gibson, James, 26
Goodheart, George, 114–16, 119
Gorski, Peter, 130
Grad, Bernard, 72
Gravity, human body and, 100
Greece (ancient)
  massage practices in, 98
  touch healing and, 58
Grey Panthers, 52

Grossman, Richard, 61, 75, 77–81
Gurney's Inn, massage practices at, 133

Haley, Alex, 14–15
Hall, Edward, 148
Handicapped
  persuasive touching and, 155
  pet touch and, 171–72
  self-touching and, 159–60
  in wheelchairs, 51–52
Hands
  energy of, 91–92
  reflexology massage of, 125
  self-touching and, 158–61
Handshakes, 145–46
  communicative touching and, 161
Hardison, James, 179
Harlow, Harry, 8–9
Harris, T. George, 138
Hausknecht, Richard, 21
Having a Caesarean Baby (Hausknecht
    and Heilman), 21
Hayashi, Chujiro, 88
Haynes, William, 138
Healing, 3. *See also* Energy medicine
    and specific therapies
  anthropology and, 10
  historical overview of, 57–62
  pet touch and, 166–68, 170
  quality of touch, importance of, 130
  self-massage for, 133
  touch used in hospitals and, 128–32
Health spas, 134
Heart. *See* Coronary care patients; Pulse
Heilman, Joan Rattner, 21
Henly, Nancy, 144, 151–52, 162, 178
Henry, Jules, 11
Heslin, Richard, 149, 153–54
High fives, 16
Hippocrates, 139
History
  of massage practices, 98–99
  reflexology practices in, 124
  of spinal manipulation, 110
Holistic medicine
  information resources on, 188
  osteopathy and, 108–9
Hollender, Marc H., 45
Homosexuality, 38–39, 176–77
Hospitals
  infant mortality in, 23–25
  touch used in, 128–32
Howell, Mary, 129
*How to Win Friends and Influence
    People* (Carnegie), 153
Hypnotism, 61

Ik tribe, 12
India, customs of, 17

Infants. *See* Babies
Ingham, Eunice D., 124
Inkeles, Gordon, 122
Intellectual development of babies, 31–32
International Center for the Disabled, rehabilitative massage practices of, 127–28
Israel, 15
Italy, 17–18

Janov, Arthur, 97
Japan
  faith healing in, 141
  Reiki practices in, 87–88
  Shiatsu practices in, 64–65, 84–85
Jesus, touch healing and, 59
John E. Fetzer Energy Medicine Research Institute, 65
Jones, Franklin, 103
Jones, Stanley, 178
Jourard, Sidney, 18–19
Joy (Schutz), 160
Joy of Touch, The (Rueger), 92
Jungle People (Henry), 11

Kaingang tribe, 11
Katcher, Aaron H., 35, 167–68, 171
Katz, Alex, 6
Keller, Helen, 56, 164
Kennedy, Adele P., 40, 43, 177
Kennell, John H., 31–32
Kenny, Elizabeth, 130
Kinesiology. *See* Applied Kinesiology
Kissing, 46–47
Klaus, Marshall H., 31–32
Knapp, Mark L., 36
Krieger, Dolores, 66–75. *See also* Therapeutic Touch
Kübler-Ross, Elisabeth, 14, 53–54
Kuhn, Maggie, 52
Kule, Sheldon, 181–83
!Kung, 13
Kunz, Dora, 71–72

Lamaze method, 21–22
  information resources on, 189
Language, touching and, 163–65
Language of the Body, The (Lowen), 96
Last Tango in Paris, 41
Lauder, Estée, 53, 154
Lauder, Leonard, 82, 146
Lawrence, D. H., 4
Laying-on of hands. *See also* Faith healing
  Episcopal church and, 59
  first scientific explanation for, 61
  at mainstream churches, 138
Leavy, Lynne, 150

Le Boyer, Frederick, 32
Lee, David, 168
Lee, Linda, 17
LeShan, Lawrence, 141–42
Levinson, Boris, 169–70
Lewenberg, Adam, 81–82
Libby, Rober W., 43–44
Ling, Pehr Henrik, 99
Littleton, Camille, 138
Lives of a Cell, The (Thomas), 62
Lourdes, 137
Love, Medicine and Miracles (Siegel), 6
Love, sex and, 42–43
Loving Hands (Le Boyer), 32
Lowen, Alexander, 96, 164
Lowenthal, Milton, 127
Lucey, Jerry, 130
Lymphatic system, Touch for Health and, 119–20
Lynch, James, 129–30, 168

McAnarney, Elizabeth R., 37
McCulloch, Michael, 171–72
MacDonald, Cese, 157
Mammals, babies of, 25, 28–29
Manipulative medicine. *See* Massage
Mann, Felix, 78
Maramus, 25
MariEl, 90
Marijuana, 182–83
Mariner, William, 12
Marshall, Lorna, 13
Mason, Adrienne, 88
Massage (manipulative medicine). *See also* Shiatsu *and specific therapies*
  Alexander technique, 102–4
  anthropology and, 30–31
  Applied Kinesiology, 113–17
  of babies, 30–33
  bioenergetics as a form of, 96–97
  Chinese, 83–84
  chiropractic, 110–13
  Feldenkrais technique, 104–7
  medical (rehabilitative), 126–28
  osteopathy, 107–9
  polarity therapy as form of, 91
  products for, 183–84
  Reflexology, 123–26
  Reichian, 93–97
  Rolfing (structural integration), 100–2
  self-massage, 132–34
  Swedish, 121–23
  Touch for Health, 118–21
Massage Book, The (Downing), 91
Masters and Johnson, 40, 44–45, 180
Masturbation, 35, 175–76
Mbuti tribe, 13
Mead, Margaret, 12

Medicine. *See also* Energy medicine;
    Scientific medicine
  Eastern and Western compared, 64–65
  holistic, 108–9
  Therapeutic Touch research and,
    72–73
*Medium, the Mystic and the Physicist,
    The* (LeShan), 141–42
Meehan, Therese Connell, 73
Men
  aging and sex, 52
  anthropology and, 14–15
  birthing process and, 20–23
  communicative touching and, 158
  handshakes of, 145–46
  pet touch and, 171
  power touching and, 144–45, 149, 152
  same-sex touching, 176–77
  sex and, 44–45
  sports touching and, 16
  touch healing and, 131
Mental retardation, 29
Meridians
  acupuncture and, 77
  Touch for Health and, 119–20
Mesmer, Franz, 61
Middle Ages
  massage practices in, 99
  touch healing and, 59–60
Milberg, I. R., 130–31
Mind/Body Effect, The (Benson), 139–41
Monkeys, experiments with, 8–9
Montagu, Ashley, 2, 11, 13–14, 17, 52,
    147, 159
  founding hospital story, 23–25
Moor-Jankowski, Jan, 9–10
Morris, Desmond, 42, 178
Moslem countries, 19
Mundugumor tribe, 13
Muscle system, Applied Kinesiology and,
    113–17
Muscular system, Touch for Health and,
    120

Naked Ape, The (Morris), 42
Namikoshi, Tokujiro, 85
National Federation of Spiritual Healers
    (Great Britain), 141
Netsilik Indians, 11
Neuromuscular system, Feldenkrais
    technique and, 105
Newborns, 23
New Testament, touch healing and, 59
Norman, Laura, 124–25
Nurses
  comforting touch of, 150
  dying patients and, 55
  power touching by, 151
  Therapeutic Touch and, 73

O'Connor, Dagmar, 45
Orgone box, 95
Osteopathy, 107–9
  information resources on, 192
Other Medicines, The (Grossman), 75

Palmer, Daniel Davie, 110–11
Parents, infant bonding with, 31
Parrish, Essie, 137–38
Pease, Allan, 145–46, 159
Pediatrics (Rosenthal), 33–34
Peebles, Rufus, 150, 156–57
Penney, Alexandra, 41
Persuasion, 152–56
Pet touch, 166–72
  elderly and, 168–69
  healing and, 166–68, 170
Pfitzenmaier, Joan, 105–6
Physiatry, 126–28
*Physical Dynamics of Character
    Structure* (Lowen), 164
Physical therapy, touch healing and, 128
Placebo effect, of faith healing, 139
Polarity therapy, 91–92
  information resources on, 191
Polio, touch healing and, 130
Posture, Feldenkrais technique and, 105
Power touching, 143–52
  affection and, 147
  handshake practices and, 145–46
  hierarchy of touch and, 144
  interpersonal space and, 148–49
  interpretations of, 150–51
  sex and, 146–48
  taboos against, 178
Prana, 63
Premature infants, 27–29
Prescott, James A., 15, 36
Primal Scream (Janov), 97
Primates
  research with, 8–10
  sexual skin of, 41–42
Prisoners, 17
  pet touch and, 168
Probst, Leonard, 162–63
Prohibitions against touching. *See* Taboos
Prudden, Bonnie, 86
Psychedelic drugs, 181
Psychoanalysis
  empathic touch and, 150
  Reichian massage and, 93–94
  touching and, 156–57
Psychology
  communicative touching and, 156–63
  persuasive touching and, 152–56
  power touching and, 143–52
Psychoneuroimmunology, 64
Psychotherapy, pet touch and, 169–70
Puberty, 35–36

Pulitzer, Ramelle, 32–33
Pulse, touch healing and, 130

Qi
  acupuncture and, 77
  definition of, 63

Reflexology, 123–26
  information resources on, 193
Rehabilitative massage, 126–28
Reich, Wilhelm, 93–97
Reichian massage, 93–97
  bioenergetics compared to, 96–97
  information resources on, 191
Reiki, 59, 87–90
  definition of, 87
  information resources on, 190–91
  recommended uses of, 90
  session described, 88–89
Reite, Martin, 9
Religion, 4
Reston, James, 78–79
Rivers, William Halse, 12–13
Rogers, Ada, 128
Rolf, Ida, 100–2
Rolfing, 3, 100–2
  information resources on, 191
Rome (ancient)
  massage practices in, 98–99
  touch healing in, 59
Roots (Haley), 14–15
Rosenbaum, Ron, 86
Rosenthal, Maurice J., 33–34, 131
Rosenthal, Robert, 131–32
Royalty, touch healing and, 60
Rueger, Russ, 92, 181
Rules against touching. *See* Taboos

Sacrament money, 60
Sado-masochism, 47, 179
Salespersons, touching by, 154
Samoa, 12
Samuels, Mike, 132–33
Schanberg, Saul, 28–29
Schutz, William C., 160
Scientific medicine
  acupuncture and, 80
  and chiropractic, attitude toward,
    112–13
  and Eastern medicine compared,
    64–65
  faith healing and, 138–41
  rehabilitative massage, 126–28
  Reichian Massage and, 95
Scott, Robert, 180–81
Self-massage, 132–34
Senegal, infant massage in, 30
Sensuality, sex and, 40–41

Sex, 5, 39–40, 182
  anthropology and, 46–47
  brain and, 42
  drugs and, 180–83
  elderly and, 52
  erogenous zones and, 46–47
  inhibitions about, 48
  love and, 42–43
  Masters and Johnson and, 40, 44–45,
    180
  masturbation, 35, 175–76
  men and, 44–45
  power touching and, 146–48
  Reichian massage and, 93–95
  sado-masochism, 47, 179
  sensual touching and, 40–41
  skin signals and, 41–42
  taboos against, 175–79
  women and, 45–46
Shamanism, 137–38
Shiatsu, 3
  definition of, 84
  information resources on, 190
  Japanese practices of, 64–65
  recommended uses of, 86–87
  session described, 85
*Shiatsu: Japanese Finger Pressure
    Therapy* (Namikoshi), 85
Siegel, Bernie S., 6, 139
Simon, Sidney, 154
Simonton, Stephanie Matthews, 54, 131
Skin
  disorders of, touch healing and, 130–31
  of elderly, 49, 51, 53
  layers of, 2
  sensitivity of, 26
  and sexual touching, 41–42
Sleeping, 16–17
Smith, Justa M., 72–73
Sokolow-Molinari, Ellen, 87–90
Solomon Islands, 12–13
Spas, 134
Speech, children and, 37
Spinal lesions, 108
Spine
  Alexander technique and, 102–3
  animal alignment of, 107
  chiropractic and, 110–13
Sports, communicative touching and,
    158
Stephenson, James, 108
Still, Andrew Taylor, 107–8
Stimulants, 180–81
Stoddard, Frederick J., 128
Stone, Randolph, 91
Stories the Feet Can Tell (Ingham), 124
Strangers, taboos against touching,
    173–75
Strauss, Murray, 43–44

Structural Integration, 100–2
Subluxations, 108, 110–11
Subways, 16–17
Sucking Power (documentary), 138
Surgery, touch healing and, 129
Swaddling, 13
Swedish massage, 99, 121–23,
    193

Taboos, 172–79
    elderly and, 177
    incest, 179
    masturbation, 175–76
    power touching and, 178
    sex and, 176–78
    strangers, 173–75
Tai Ch'i, 63
Takata, Hawayo, 88
Talbot, Fritz, 24
Tasaday tribe, 11–12
Therapeutic Touch, 3, 66–75
    and acupuncture compared, 70
    connective-tissue injuries and, 129
    description of, 55–56
    energy field self-test and, 67
    history of, 71–73
    information resources on, 189–90
    process of, 68–70
    recommended uses of, 75
    theories about, 74
Therapeutic Touch (Krieger), 67
Thie, John F., 118–21
Thomas, Lewis, 62
Tips, touching as increasing size of,
    155–56
Tobiason, Sara Jane Bradford, 49–51
Todres, David, 27
Tonga, 12
Touch for Health, 118–21
    information resources on, 192
Touch for Health (Thie), 118

*Touching, the Human Significance of the
    Skin* (Montagu), 2
Touching for Pleasure (Kennedy and
    Dean), 40, 177
Turnbull, Colin, 11

Usui, Mikao, 88

Vascular system, Touch for Health and,
    120
Vegetarianism, polarity therapy and, 92
Verny, Thomas, 21

Waitresses, 155–56
Way to Vibrant Health, A (Lowen), 97
Weather, touching and, 182
Weil, Andrew, 139
Well Body Book, The (Samuels), 132–33
West, Mae, 43
Westheimer, Ruth, 48
Wetzel, Christopher G., 155
Wheelchairs, elderly and, 51–52
Whitcher, Sheryle J., 129
Whitehead, William E., 130
Winter, William, 172
Wisconsin Regional Primate Center, 8–9
Women
    aging and sex, 52
    birthing process and, 20–23
    handshakes of, 145
    massage during childbirth, 21–22, 30
    power touching and, 147–48, 151–52
    same-sex touching, 176–77
    sex and, 45–46
    touch healing and, 129
World Federation of Healing, 141

Yarborough, Elaine, 178
Youngest Science, The (Thomas), 62

Zone therapy, 123–26